THE RIGGING HANDBOOK

ALSO AVAILABLE IN THE SHERIDAN HOUSE GUIDES SERIES

Boat Buying Basics

Marine Electrical and Electronics Bible: A Practical Handbook for Cruising Sailors, 4th Edition

The Rigging Handbook: A Comprehensive Guide to Sailboat Rigging

Understanding Boat Batteries and Battery Charging: 2nd Edition

Understanding Boat Plumbing and Water Systems: 2nd Edition

Understanding Boat Wiring: 2nd Edition

Understanding Marine Diesels: 2nd Edition

THE RIGGING HANDBOOK

A COMPREHENSIVE GUIDE TO SAILBOAT RIGGING

HERB BENAVENT, THE RIGGING DOCTOR
FOREWORD BY LIN PARDEY

Essex, Connecticut

An imprint of The Globe Pequot Publishing Group, Inc.
64 South Main St.
Essex, CT 06426
www.globepequot.com

Copyright © 2026 by Herb Benavent
Foreword copyright © 2026 by Lin Pardey
Unless otherwise noted, all photos and illustrations copyright © 2026 by Herb and Maddie Benavent

All rights reserved. No part of this book may be reproduced in any form or by any electronic or mechanical means, including information storage and retrieval systems, without written permission from the publisher, except by a reviewer who may quote passages in a review.

British Library Cataloguing in Publication Information available

Library of Congress Cataloging-in-Publication Data available
ISBN 9781493086009 (paperback) | ISBN 9781493086016 (epub)

Limit of Liability / Disclaimer of Warranty: The author and The Globe Pequot Publishing Group, Inc. (GPPG) expressly disclaim any and all liability for the use of any materials, information, or methods described in this book, including, without limitation, for accidents, injuries, illnesses, damages, or death sustained by readers who engage in the activities described or promoted herein. No representations or warranties are made as to the accuracy or completeness of the contents of this work, and all warranties, express or implied—including, without limitation, warranties of fitness for a particular purpose —are expressly disclaimed. The opinions presented are solely those of the author and are provided for informational purposes only.

Endorsement Disclaimer: Reference to any individual, organization, website, or other resource in this book does not constitute an endorsement by the author or GPPG of such party or the information, products, or services they may provide, now or in the future.

Contents

Foreword vii
Introduction 1

Part I: FOUNDATIONS OF RIGGING 3
Chapter 1: Looking Up 5
Chapter 2: Setting Sail 13
Chapter 3: Rigging It 21
Chapter 4: The Mast Itself 31
Chapter 5: How Running Rigging Ties In 55
Chapter 6: How Sails Make It All Work 75
Chapter 7: Going to the Top 93
Chapter 8: Down to the Bottom 97
Chapter 9: Rigging Materials 113
Chapter 10: Building It 121

Part II: MECHANICS OF RIGGING 163
Chapter 11: When Things Go Wrong 165
Chapter 12: How to Carry Out a Rig Inspection 185
Chapter 13: When to Replace 199
Chapter 14: Avoid Repeating Your Mistakes 205
Chapter 15: Steel versus Synthetic 211
Chapter 16: A Frayed Knot 221
Chapter 17: Setting Up the Standing Rigging 237
Chapter 18: Setting Up the Running Rigging 251
Chapter 19: Tune Your Sailboat .283

Afterword 297
Glossary 299
Index 311

Foreword to *The Rigging Handbook*

Finally they had the chance they'd been fighting for. *Steinlager 2*'s crew, led by the late Sir Peter Blake, had driven their boat to line honors in five legs of the 1989 Whitbread Round the World Race. Now they were on their way across the final ocean that lay between them and a total victory. Peter glowed as he recalled the excitement of being decisively in the lead as they headed out into the Atlantic toward Southampton, England. "We finally grabbed that trophy because we had a great boat, great crew, and one real bit of luck," he told me.

This had been Peter's fourth attempt at winning what was, in 1989, considered the most difficult ocean race in the world. For 12 years this race had dominated Peter's life. Every one of his crew were long-term friends who shared Peter's passion. But the real bit of luck? Just a few nights after departing Fort Lauderdale, Peter had come on deck for the start of his watch. Following the routine he'd established from the very first, he put on a head torch and took a walk around the deck. The minute he returned to the cockpit he called the previous watch back on deck. There were a few grumbles, as most were already climbing into their bunks. He beckoned for them to head onto the foredeck, where he pointed out a frayed pull line he had spotted on one of the halyard snap shackles. In tones he later admitted were definitely accusatory, he asked, "Why wasn't this reported?" He then pedantically reminded the whole crew that, at the start of every watch, day or night, every detail of the deck gear and rigging had to be inspected and any change at all reported to him.

Two days later, Peter retired below after a boisterous watch. The boat was boldly forging to windward in brisk winds. "Peter, probably nothing. But come on up and look," called the captain of the new watch. "I spotted a small bit of chipped paint on the edge of one of the chainplates."

Peter didn't hesitate. "Bring her about," he called, even before he climbed into the cockpit. Further inspection revealed that under the chipped paint, the chainplate had a definite crack right next to the clevis pin holding the main shroud. With a jury-rigged shroud and cautious sailing, *Steinlager 2* carried on to finish the final leg and take top honors for the whole race. Peter called it luck; I call it good seamanship.

FOREWORD

The moment I began reading *The Rigging Handbook*, Peter's story leapt back into my mind. I felt honored when Herb asked me to write a foreword to this book. At first, I thought the timing just wouldn't work for me, as my partner, David, and I were heavily engaged in preparations to set sail for Australia. I was writing an email to offer my apologies when I took one more look at the table of contents and came across something that I definitely wanted to know—what are the drawbacks versus the advantages of substituting synthetic rope for stainless steel wire? I rationalized adding one more complication to my life by saying, "I don't have to read every page of this manuscript. I can skim it, read a few of the sections that interest me most, then explain to beginner or intermediate sailors why they should part with some of their 'freedom chips' to buy this book."

Within a few pages I was no longer skimming. Instead, I lay in bed late at night reading every page and making mental notes. And as I followed along with Herb's logical presentation of the rhyme and reason for each wire, each rope on board, and how they worked on each type of rig—from the simplest and most ancient of sailing rigs right up to the most complicated modern high-performance and cruising adaptations—I was reminded again why, despite being fully involved for six decades (and counting), I have never gotten bored with a sailing life. There is always something new to learn, or sometimes to relearn.

Herb presents a careful and readable analysis of rigs and rigging materials, and a comparison between the most modern setups and their more traditional counterparts. He emphasizes the underlying simplicity of the components that make up a sailing rig and the way they silently do very complex jobs.

As he dissects and makes rigging more understandable, he also explains how you can build or repair your own rig even in the most remote of locations, using the simplest of tools. And by doing so, Herb encourages readers to take a truly "hands-on" approach to tuning and maintaining their own rig. But best of all, this book shows why it is important to inspect each component of your rig, not only during a preseason, fully dedicated rig inspection but also on a casual, day-to-day basis.

As I read *The Rigging Handbook*, I shared occasional tidbits with David. Then, a few days ago, I shared my first draft of Peter Blake's "good luck" story as I wrote it for this foreword. That same day, just before dark, David took a stroll forward to check the snubber line on the anchor chain. As he approached the anchor roller, David noticed the shackle attaching the tack of the headsail to the roller furler drum had worked loose. The shackle pin was hanging on by just one turn of the thread. The whole furling system had been removed, serviced, and reinstalled 16 months previously by a well-respected rigging team. I'd watched them dip the shackle pin in Stay-Loc to keep it solidly in place before they tightened it with an Allen wrench. How long had the shackle pin been working its way free? Not sure, because like most roller

FOREWORD

furling gear, ours just quietly went along doing its job, so the only time we inspected it was when we were doing a predeparture rigging check. During the four months since that last inspection, we'd crossed the very boisterous Tasman and been cruising up the Queensland coast of Australia. In our case, we wouldn't have lost the rig or a race if that shackle pin had gone overboard. But had this happened underway, we might have been unable to use the sail until we encountered calm seas. Fortunately, the morning after David's discovery there was absolutely no wind. So, we were able to unfurl the sail, reconnect and secure the shackle, and reflect on the fact that reading this book probably contributed to our "luck."

Though not a major incident, our own small rigging mishap did sum up the point of this book and also this foreword. Understanding your rig and all its components, and knowing how to inspect and maintain it, is part and parcel of good seamanship. It can lead not only to better sailing performance, but also encourage you to use your own skills and avoid the need to call on the limited supply of luck gifted to all of us who venture offshore under sail.

Lin Pardey
Voyager, US Sailing Hall of Fame inductee, and author of 12 books
on cruising and seamanship, including *Storm Tactics Handbook*

Introduction

Rigger: n. a person who rigs or attends to the rigging of a sailing ship.

The goal of this book is to teach you how to become your own rigger so that you might attend to the rigging of your own sailing vessel. Whether you want to be able to rerig an entire boat, or just know how to monitor the condition of your own rigging, this book can be your guide. Understanding how rigging works is also vital to safety while sailing.

Within these chapters, you will learn about all sorts of different materials that are available for rigging today, as well as how to select the right material for the job at hand. More importantly, you will learn how and why parts of your rigging fail, how to identify points of weakness and failure, and how to design a replacement that is better suited to your boat. I will teach you not only how it works, but also how to problem-solve in unique situations. A manual tells you how to do something—my hope with this book is that it will teach you how to figure it out for yourself.

There is no one path of formal schooling in rigging. Most professional riggers simply learn on the job, where experience is their teacher. After years aloft, they have seen it all, learned what works and what doesn't, and figured out how to get the job done quickly and safely. This wealth of experience earns them the rank of Master Rigger—and it takes years to achieve such a title.

The other way to learn how to become a rigger is to study the physics of a sailboat. You can build a solid foundational knowledge and calculate what the actual loads are going to be, and then work through the material data to figure out what will work best for your specific boat.

Knowledge can be acquired over time through practice or transmitted quickly through literature. The fact that you have picked up this book means that you, like me, appreciate learning things from books and then applying that knowledge to the real world around you.

Unlike most riggers, my career started in dental school. Dentistry obviously focuses on teeth, but what most people don't realize is that dentists also need to learn how to be engineers. Designing a bridge that supports cars is not all that different from designing a bridge that supports chewing. You need to learn about the properties

INTRODUCTION

of different materials and figure out the minimum thicknesses allowed in order to build something that will withstand the planned forces for the life of the bridge.

During dental school, I spent all my free time reading about sailing because I wanted to cross an ocean after I graduated. I bought an old classic boat that needed to be completely rebuilt, which allowed me to apply all of the knowledge I had acquired through reading over the years. In the early 2010s, synthetic rigging was still a new and rather unconventional material, definitely not a mainstream name like it is today. I was determined to rig my 45-foot cutter with synthetic standing rigging and deadeyes because I believed it could have more longevity than steel. I was also excited by the challenge because at that point in time, there was no effective method to do this for sale at the local chandlery. I had to invent it myself. I invented the shroud frapping knot and developed the grommet splice that you will read about later, which made this undertaking possible—all by using the data specs that are available for these materials. By calculating the forces involved, how they would affect rigging creep, and sizing everything accordingly, I was able to build the first of its kind: a fully synthetic standing-rigged sailboat.

As a test for our new rig, my wife and I sailed from the United States to Portugal. I was satisfied with the results and shared my findings with the rest of the world for free on my website and YouTube channel, @RiggingDoctor. After 18,000 nautical miles and 5 years of full-time cruising, I feel this method is well tried and sound for others to replicate, if you so choose.

I have written this book to teach you what I have learned about the fascinating science and art of rigging in order to enable you to be confident in fixing your own boat. This ability will not only save you the money you would have spent on hiring a rigger, but also provide the peace of mind and satisfaction of being self-sufficient on your sailboat.

I hope you enjoy the journey!

Note: While reading this book, it is always important to remember that the cause of an accident is an unplanned or uncalculated event. To avoid accidents with your rigging work, you always want to calculate a little extra to act as a safety margin and protect you against those unanticipated moments. This extra margin is known as the safety factor and ranges from 1.5 to 20, depending on the material and application. When you calculate the minimum size for your rig, you always want to include the safety factor in there as well so that your setup is built to withstand the unknowns that will occur.

In the tables that I provide later in the book for chainplate and bolt sizing, these safety factors are already incorporated, allowing you to simply look up what you need and saving you a lot of calculating.

PART I
Foundations of Rigging

CHAPTER ONE

Looking Up

When looking up at the mast, it might seem a little overwhelming with all the rigging tangled up in the sky. The lines seem to go in every direction and grab onto the mast in all different places, coalescing into a giant spider web in the sky.

Rigging has one simple job to do: keep the mast up to hold sails in the wind. An easy way to remember this is: The stays make the mast stay put. They do this by pulling, never by pushing. If you have ever held a blanket or towel up on a windy day, and had the wind fill the blanket and felt the pull, then you have experienced what it is like to be a mast! As the wind fills the blanket, it tries to pull you over and you have to resist this action. Imagine if you had a friend holding your shoulders to keep you from falling over; in this situation, you would be the mast and your friend would be the stay.

While it might look complicated at first, rigging is actually very simple. It can be separated into two categories: standing rigging and running rigging. Standing rigging is the rigging that makes the mast stand up, while running rigging is used to make the sails work. On most boats, if there is a rope that you can pull to tighten or loosen, this would be part of the running rigging; while if it is a piece of rigging that is either tied or locked into a particular position or length, it is safe to say that this is a piece of standing rigging. Running rigging will be discussed further in a later chapter. First, let's learn about the parts of the standing rigging.

The mast is standing high in the air, coming straight out of the deck of the boat. Any pole that is standing up in the air like that will try to fall over, so to keep it standing up, it needs some support. A mast has four main stays that keep it from falling forward, backward, to the left, or to the right. These stays run from the top of the mast to the bow, to the stern, and to the port and starboard sides of the boat.

Any stay that runs forward or backward will have the word "stay" in its name, while any stay that runs to the sides of the boat (also called athwartship) will be called a shroud. The term shroud comes from old tall ships that used organic fibers for their rigging. These fibers were not very strong, so they in turn used a lot of them, all working together. There were so many that it looked like a shroud of clothing wrapping around the mast. With modern fibers, we no longer need to have such duplicity with stays and can have one single stay that will take the entire load. As a result, modern rigs look much less intimidating than historical rigs, but it can still seem like a lot if you don't know what you are looking at.

BACKSTAY

The backstay runs from the back of the boat to the top of the mast and keeps the mast from falling forward. While it might be easy to think that a stay holds the mast up by pushing it into that position, it is important to remember that any piece of rigging only pulls, never pushes. The backstay does nothing to keep the mast from falling backwards; instead it pulls the top of the mast back and keeps the mast from falling forward. The importance of this stay will be discussed further in chapter 5, where we will learn about luff tension on a sail and how it affects the boat's ability to sail upwind. For now, just remember that the backstay pulls the mast backwards, hence its name, and plays directly into the stay that pulls in the opposite direction.

HEADSTAY

The headstay! This is the stay that everyone is concerned about. The headstay runs from the tip of the bow of the boat all the way to the top of the mast. The reason it is called the headstay is because the tip of the front edge of the bow is called the stem of a boat, and the top of the stem is called the stemhead. The top of the mast is also called the masthead. This stay runs from the masthead to the stemhead, hence the name headstay.

The headstay has a simple job: All it really needs to do is hold the mast up and keep it from falling backwards. But sailors have found ways to further complicate its job by attaching sails to it, which are called headsails and will be discussed further in chapter 6. The headsail will pull on the headstay in a sideways manner, which means that the top of the mast will also be pulled to the side! The tighter the headstay is, the better it can remain straight as the headsail pulls on it, but this is a lot of side load being put on one piece of rigging that is only attached at the ends. The amount of deflection to the side is called "headstay sag"; this will be discussed in greater detail in chapter 5, where you will learn how the sagging of the headstay can make the headsail generate more power to pull the boat even faster through the waves.

To control the tension in the headstay, you could easily tighten the headstay itself. If you make it shorter, then you would make the headstay tighter, which would reduce the amount of headstay sag. This is all well and good, until you realize that the sailmaker made the headsail to fit the length of the headstay, and if you make the headstay shorter, suddenly the headsail will be too long for the stay. To circumvent this issue, headstay tension is easily adjusted by tightening or loosening the backstay. Remember, the backstay pulls the mast backwards, which in turn makes the headstay tighter without changing its length. On some boats, the backstay is easily adjusted via complicated, pre-installed mechanisms that allow such adjustments to be readily made. On most boats, this complexity is negated to keep the rigging mechanisms simpler, and there is no way to quickly and easily adjust the backstay tension without busting out a bunch of tools to get the job done.

SHROUDS

The backstay only pulls the mast backwards and helps the headstay maintain its tension to limit headstay sag, but these two stays alone would just flop over with the mast in any form of wind. To keep them upright, the mast needs stays that go to the side, providing some form of lateral resistance. All the stays that go to the side of the boat are called shrouds, and they are aptly named based on where they attach.

Cap shrouds: The highest of the shrouds attach at the head of the mast, but to not use the word *head* again, riggers went with a different word that follows in the same vein: *cap*, from the word *capital*. The shrouds that go to the top of the mast are the cap shrouds.

Lower shrouds: The shrouds that go to the lowest point of the mast are the lower shrouds. If you have any other shrouds between the cap shrouds and the lower shrouds, they will be aptly called intermediate shrouds. In chapter 4 we will delve deeper into the intermediate shrouds when we learn about different mast setups.

The lower shrouds play a special function in addition to supporting the middle section of the mast. They are paired, and one pair runs forward while the other pair runs backward. The pair that runs forward is called the *forward lower*, while the pair that runs backward is called the *aft lower*.

These, as you can imagine, help hold the middle section of the mast in a side-to-side fashion, but also in a front-to-back fashion. An easy way to think about it is that the top of the mast is held in place by four stays running front to back and left to right. These form a cross when viewed from above. When you get down to the midsection of the mast, the four lower shrouds do the same thing, but they are rotated roughly 45°, so each one pulls towards the corner. Instead of making a cross, they make an X. Their function will be further discussed in chapter 4 and chapter 10.

The headstay, backstay, cap shrouds, and lower shrouds are the basic stays that you will see on most boats. However, that doesn't mean that these are the only stays you will encounter, as there are many extra

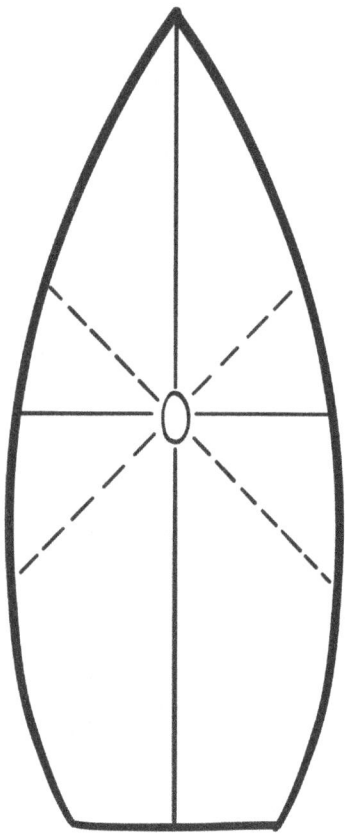

The top stays form a cross while the lower stays form an X; together they support the mast in all directions.

LOOKING UP

pieces of rigging that have been designed and implemented over the past few thousand years of sailing.

OTHER STAYS

Some other stays that you will see make a diamond outline on the mast, and are aptly called diamond stays. They are often seen in situations where you need the effects of lower or intermediate shrouds, but simply can't rig them because of some kind of limitation. For instance, the mast might need to be able to rotate, or the mast is very thin and needs a little extra help to keep it in place, hence you don't want the added weight or wind drag of having the stays run all the way down to the deck. Diamond stays help provide rigidity to the mast and prevent it from bending with as little additional rigging as possible.

Diamond stays can also be classified as a pair of diagonal stays, since a diagonal stay is any stay that runs a diagonal path on the mast. Diagonal stays are commonly seen on "linked rigs" where the stays are broken up into many shorter pieces running from one place to another on the mast. In a linked rig, the cap shroud and intermediate shroud would be shared by a single cable that runs from the deck to the first spreader. The spreader tip would then have two additional stays start there, a diagonal stay that would run up at an angle to the base of the second spreader, and a cap shroud would keep running vertically to the tip of the second spreader and then up to the top of the mast.

However, "continuous rigs" are rigs where the stay runs continuously from the top of its attachment all the way down to the deck of the boat. In a continuous rig, there would really never be a diagonal stay, since it would keep running vertically to get to the deck. This might sound complicated right now, but it will all make sense in chapter 4 when we learn about different mast types.

Another common set of stays are what look like an extra set of headstays and backstays set

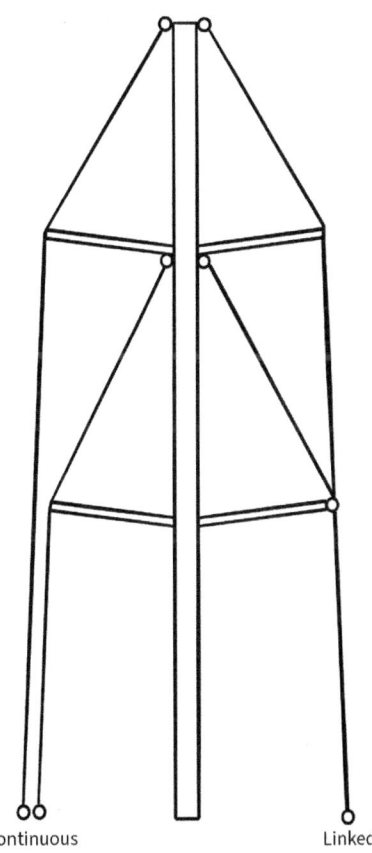

Continuous   Linked

In a linked rig, the entire length of the diagonal stay is contained between the spreader tip of the lower spreader to the spreader base of the upper spreader. In a continuous rig, the stays run all the way down to the deck.

about three-quarters of the way up the mast. These are seen on cutter sailboats (see chapter 3). These additional stays simply exist to hold additional sails in the wind.

MULTI-MAST SETUPS

On sailboats with more than one mast, each stay will have its name changed slightly to reflect its role on that specific mast. The forwardmost and tallest mast on a sailboat is called the mainmast, and as a result, every stay will have "main" added in front of or replacing part of its own name.

The mainmast stays would be:

- Mainstay instead of headstay
- Mainbackstay instead of backstay
- Main cap shroud instead of cap shroud
- Main lower shroud instead of lower shroud

The second and smaller mast, which comes behind the mainmast, is called the mizzen, and the same rules apply.

The mizzen stays would be:

- Mizzenstay instead of headstay
- Mizzenbackstay instead of backstay
- Mizzen cap shroud instead of cap shroud
- Mizzen lower shroud instead of lower shroud

Now, when you have two masts, there exists the opportunity to have a stay that runs between the two masts to give each one extra support. This is called the triatic stay.

On a schooner, where the forward mast is shorter than the aft mast, the names are again different. Since the mainmast is always the tallest mast, this means that the mainmast is in the back and the forward mast is aptly named the foremast. Once again, all the names are changed to include "fore" in them.

The foremast stays would be:

- Forestay instead of headstay
- Forebackstay instead of backstay
- Fore cap shroud instead of cap shroud
- Fore lower shroud instead of lower shroud

LOOKING UP

You might have noticed that regardless of the type of boat, a lot of people when talking about non-schooners refer to the *headstay* as the *forestay*, and the respective sails as the foresails, but this is a common mistake. If there is only one mast, then that single mast is the mainmast and all the sails and stays are respective to that nomenclature. This is why the mainsail is called such—it is the sail attached to the mainmast!

All of these details will be discussed further in chapter 3 when we get into the fun differences between the different rig configurations. For now, you know the different stays and what they are called, as well as why they are there. These stays are all part of the standing rigging on a sailboat. The next chapter will teach you all the parts of the boat that are referred to as running rigging.

CHAPTER TWO

Setting Sail

While standing rigging is mostly static, running rigging is always being changed to manipulate the sails. The main categories for running rigging are *halyards*, *lifts*, *hauls*, and *sheets*. These lines will have the same function on any boat, regardless of their placement, size, or style.

In general, a halyard raises a sail, a lift raises a pole, a haul pulls a sail back, and a sheet trims the sail to harness the power of the wind. Just like with the standing rigging, where each stay is named for the location on the mast where it attaches, each piece of running rigging is named based on what it does and which sail it moves. For example, the main halyard raises the mainsail, while the jib sheet trims the jib.

It is important to know that sails make power thanks to their curves. A flat sail doesn't generate as much lift as a sail with a nice curve to it. The bulge that appears in the sail is called the "belly" by sailors, and the "draft" by engineers. The size and position of the draft are controlled by the running rigging and directly relate to the amount of power the sail will generate.

HALYARD

In the simplest of forms, the halyard raises the sail. The main objective of this piece of running rigging is to get the sail into the wind! These lines are easy to identify because they will go up the mast. If your halyards terminate at the mast, you will see them all coiled up on the mast winches; but if the halyards are led aft to the cockpit, then they will contribute to what is colloquially referred to as "cockpit spaghetti," which is nothing more than all the running rigging tossed loosely into the cockpit, where it becomes a confusing jumbled mess.

While the first task of a halyard is simple enough to comprehend, the second is a little less obvious at first. Tension on the halyard will move the belly of the sail forward. To generate the most power possible, the draft should be centered at the middle of the sail, front to back. If the draft is farther forward, the sail will generate less power, and if the draft is farther aft, the sail will generate less power but also more drag, which will cause the boat to heel over more than it should.

In light winds, easing the halyard will ease the luff tension and move the draft aft. As wind speed increases, the sail will generate more power; in order to control that power by reducing the sail's efficiency, you can add tension to the halyard, which will tension the luff of the sail and move the draft forward.

An important note about headsail halyards with roller furling is that they should never run parallel to the stay as they risk becoming twisted around the stay in what is termed halyard wrap. This is a very dangerous situation that can lead to breaking the

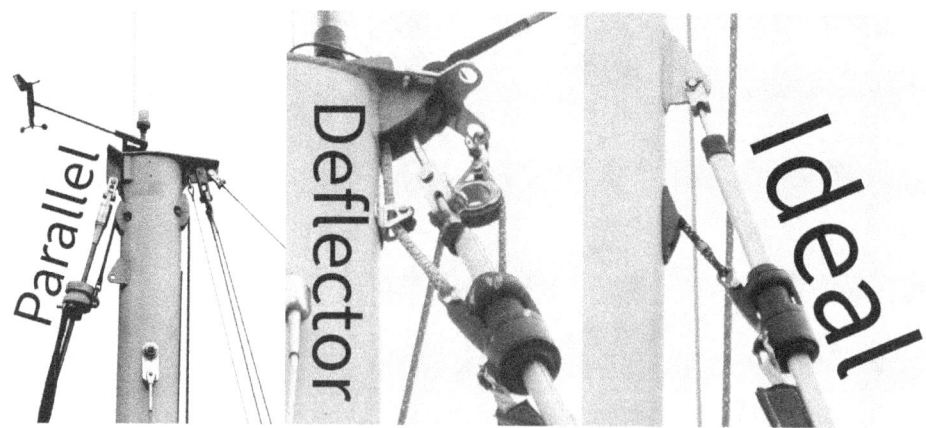

SETTING SAIL

furler, the stay, and the halyard! To avoid this, the halyard is either designed to exit the mast a good bit lower than the stay or employing a deflector to create the difference in angle is all that is needed to avoid disaster.

On a boat, you will rarely find a line whose only name is *halyard*. They all have additional names, just like all the other parts of the rigging. The halyard will be named based on the sail that it manipulates.

The mainsail is raised by the main halyard.

The mizzen sail is raised by the mizzen halyard.

The genoa or jib is raised by the genoa halyard or jib halyard.

The staysail is raised by the staysail halyard.

The spinnaker is raised by the spinnaker halyard.

> Just remember, the halyard raises the sail, and halyard tension will adjust the position of the draft.

LIFT

While halyards lift the sail, *lifts* lift the spars. The most common lift you will find on a boat is the topping lift, which holds the boom up when the mainsail is lowered. It is easily identified as the line that runs from the top of the mast to the end of the boom.*

While this line is often forgotten, it offers some special sail trimming that is very useful for downwind sailing in tight quarters and will be discussed further in chapter 5.

HAUL

In its simplest form, a haul is a line that pulls a sail. The name added before the word "haul" will tell you how it will pull the sail. The most common type of haul is the outhaul, which from its name, you can presume pulls a sail "out."

The outhaul is located on the boom of the main or mizzen sail and it pulls the clew of the sail out to the end of the boom. The outhaul does a wonderful job of controlling how deep the draft of the sail will be. If the outhaul is pulled tightly, the sail will be pulled flat to the end of the boom and therefore will not be able to bulge out very much. This is great for depowering the sail as wind speeds increase, as a flatter sail makes less power. Doing the opposite is advantageous for light air sailing, where

---

* While most boats have a topping lift, it is not going to be found on a boat with a rigid vang, as the rigid vang does the job of keeping the boom from falling when the mainsail is lowered.

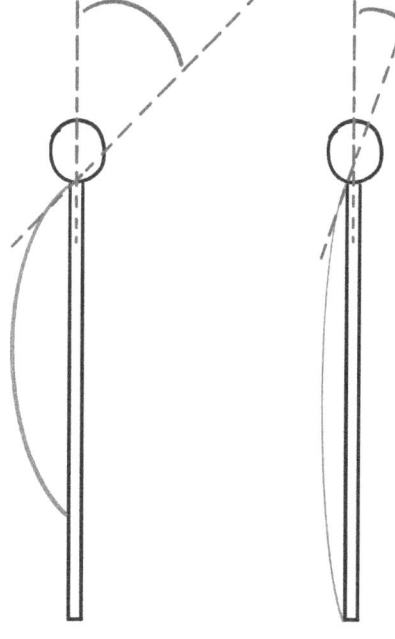

easing the outhaul will let the sail generate a deeper draft and more power.

While we all want to trim our sails to generate the most amount of power possible, it is important to also trim for wind angles. When going upwind, the angle of attack of the sail limits how far upwind you can point the boat. Having a large draft increases the angle of attack and while the sail generates more power, it can't sail too far upwind. By tightening the outhaul, the angle of attack will be decreased as the sail flattens out, allowing the sail to point farther upwind.

In summary, the outhaul controls the depth of the draft, which can power or depower the sail. Likewise, the halyard can move the location of the draft fore or aft by increasing or decreasing luff tension.

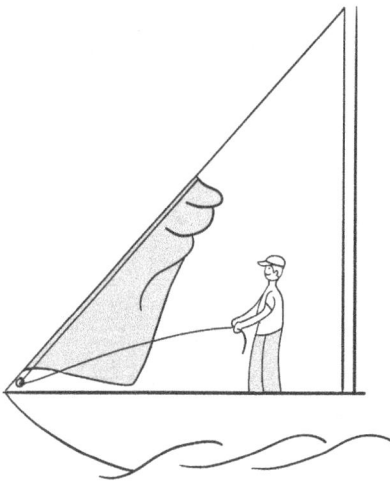

A less popular but vitally important type of haul is a downhaul. This line is critical if you need to drop a sail in a hurry! The downhaul attaches to the head of the sail and comes down to the deck. If you release the halyard in really high winds, the wind pressure on the sail might not let it slide down, and now you have to pull on a flogging cloth while being tossed around in high winds. Having a downhaul rigged up makes this situation go a lot smoother. Simply release the halyard and if the sail won't come down, give the downhaul a tug. As the sail begins to come down, it will get smaller and the wind pressure won't be able to keep holding the sail up; as a result, the sail will come crashing down at a speed that will make everyone on board very happy.

The downhaul is very important with hank-on headsails, because if they don't want to come down, you need to climb out onto the tip of the bow to start wrestling the sail down. The downhaul would simply run through a turning block at the tack of the headsail and lead back to the location of the halyard. Now, you can release the halyard and pull the headsail down in a hurry without risking your life!

SHEET

We have all heard the expression "three sheets to the wind." This is actually a nautical term derived from tall ships that had three headsails. If the crew was drunk and lost control of the sheets, the sheets would be flapping in the breeze! Fun little fact, but what exactly is a sheet?

The sheets are the lines that connect to the aft corner of the sail called the clew. They control the position of the sail relative to the wind and are the most important piece of running rigging. In chapter 16 you will learn about the different types of ropes that exist so that you can select the right kind for your boat.

As with all the other pieces of running rigging, they are named for the sail they control: jib sheets for the jib, spinnaker sheets for the spinnaker, mainsheet for the mainsail. The sheets will be under considerable load, which is why they also have their own winches in the cockpit, or some form of block and tackle system to give you the mechanical advantage to trim the sheet when the wind builds.

In chapter 5, we will discuss reefing, but a simple way to know that "it's time to reef" is when the sheet becomes too heavy to handle. If you can't pull the sheet in to trim the sail, that is a nice reminder that you are exceeding what the naval architect decided were the maximum loads your boat should experience. It doesn't mean you need bigger winches to trim your sails—it means you need to make your sails smaller, via reefing, to reduce the loads.

SPECIAL LINES

While the rigging mentioned so far is the mandatory minimum to rig a boat, there are always more lines that make sailing better and safer.

Guys: These are lines used to control the position of a spar, and on contemporary boats are used almost exclusively to control the position of the spinnaker pole for spinnaker sailing or whisker pole for poling out a headsail. As with the other forms of running rigging, their name is modified by what they do or where they go, giving you names like "working guy" or "after guy."

Cunningham: Halyard tension and outhaul tension control the size and position of the draft of the sail, but you can only get so much tension with the halyard before your arms give out. Briggs Cunningham figured out that if you attach a block and tackle about a foot or so above the tack, you can create a ton of luff tension with ease and perfectly tune the position of the draft. The Cunningham is a special type of downhaul fitted to most boats whose owners care about optimizing sail performance.

Vang: The vang is another block and tackle setup that is found under the boom, running at an angle between the mast and boom. This structure limits how far up the boom can move up. This is important for downwind performance, as the mainsheet offers no downward pull and the boom can rise up as wind pressure builds on the mainsail. As the boom rises, the mainsail twists and the sail spills its wind against the wishes of the crew. Vangs are rather commonplace on all performance-oriented boats.

Traveler: The traveler is a fancy contraption that controls the position of the mainsheet. It consists of a track and car that moves the mainsheet block athwartship. This means that you can control not only the tension in the mainsheet to trim the mainsail, but also where the mainsheet pulls from. The traveler will have its own block and tackle system to pull and adjust this position. This is normally the smaller line located near the mainsheet.

Lazy Jack: These are a cascading netting of lines hanging from the mast and attaching to the boom. Their job is simple: They catch the mainsail when it is lowered. You can sail without lazy jacks, but when you drop your sail, it will likely fall onto the deck. Lazy jacks simply make life a little easier by holding the sail on the boom when it comes down.

Jackline: Jacklines are a valuable safety line that runs from bow to stern on the deck. Their purpose is to give you a secure place on which to clip your tether to keep you attached to the boat should you fall overboard. The statistics for survival when swept off a boat are dismal, so the best way to ensure your safe arrival at your next shore is to stay attached to the boat. Obviously this still isn't a guarantee of survival, but it (hopefully) lets the other crew members reach you to pull you back on board.

Jacklines should be permanently attached to the bow and stern cleats and run in a way so that they go *over* all other bits of rigging. The goal is to clip into the jackline while you are still in the cockpit and walk all the way up to the tip of the bow without needing to unclip and reclip to get around an obstacle or snag.

Lifeline: Lifelines are the little fence lines that run around a sailboat. They are set very low and are more of a tripping hazard than a safety feature if you are standing up. This is why in a storm, you want to crawl on the deck and stay lower than the lifelines. If you fall, hopefully you will catch the lifelines and prevent yourself from falling overboard. Unfortunately, the little lifeline stanchions are not very strong, and if you actually did fall into them, they would bend out of the way to help you continue on your journey off the deck. While people often think lifelines provide a form of safety, their real purpose is to help hold lowered sails on the deck and keep them from washing into the waves when they are stowed on the deck.

In addition to the regular lifelines, it is a good idea to also rig chest-high lifelines, which run from the stern rail up to the shrouds. While regular lifelines offer a false sense of security, chest-high lifelines actually do provide some security from falling overboard. You can confidently clip your tether to them and walk forward to the mast without any concerns of getting snagged on anything on your way forward; and if you do fall, they are at the right height to help catch your balance again without falling overboard.

SETTING SAIL

REEF LINE

When winds build in strength, reef lines make it possible to tie your sails into a smaller size, which reduces the amount of power they produce as well as the stress on your rigging. Reef lines attach along the leech of the sail and will become the new clew of the reefed sail. In chapter 5 you will learn the proper way to rig your reef lines as well as a nifty alternate way to rig them.

Reef lines should be permanently rigged so that when winds build, you just need to use them—not start setting them up. On most boats, the reef lines will just run through the leech of the sail at the various reefing points, but some boats will also have a reefing line setup for the tack of the sail as well. For boats with tack reefing lines, they can come in two flavors: single or double.

Single reef-line setups use a single line that runs through both the tack and clew reef points. The advantage of this setup is that you only have one line to tighten to reef the sail. Double reef-line setups have two lines to pull: one for the tack and a second one for the clew.

While the single reef line seems simpler, the line has more turns to take, which leads to greater friction and a lot more line to pull. It also doesn't give you the ability to trim the tack or clew independently. You pretty much just tighten it down as far as you can and call it a day, whether that is good for the sail trim or not.

Double reef-line setups let you trim the tack and clew independently. This lets you make the tack very tight to preserve the sail and prevent rips in the luff, while also letting you control the clew position and tension. Trimming the clew is critical if you are sailing on a close reach or on a broad reach. For a close reach in high winds, you will want the clew tight to flatten the sail, since the reef line is doing the function of the outhaul and clew line, pulling the sail both back and down. If you are on a board reach, you might want a smaller but fuller sail, in which case easing the clew line will give you a nice deep draft to generate as much power as you can comfortably from the smaller sail in high winds.

CHAPTER THREE

Rigging It

While all sailboats work by using sails to propel them, the different types of rigging to support the sails can vary wildly in style and function. The silhouette of the rigging, as seen from a distance, is what best characterizes different sailboats, and as a result is how each one is classified today.

Let's briefly go back to the very beginning, where it all started. Sailing is an ancient human activity that we have improved upon over the past several thousand years, yet we still employ the same basic principles.

It is important to note that sailing started in various parts of the world around the same time, and each ancient civilization had slightly different ways of accomplishing the same task. The Vikings started sailing everywhere during the Bronze Age, somewhere between 4000 and 2300 BCE. The Egyptians decorated a vase with a ship under sail around the year 3500 BCE. The Polynesians were sailing far and wide across the Pacific Ocean around the year 1000 BCE.

The first sailing ships of the Vikings and the early civilizations living along the Fertile Crescent all used very similar sail plans that would pull the boats with the wind. They had a yard hung from a mast and a square sail that hung from the yard. The Polynesians had a second yard attached below the sail as well, which gives this sail the name "crab claw sail." These simple sail setups enabled trade, as people and goods were able to travel great distances over water. Square sails were not perfect though, and only really worked for going downwind. If you needed to travel upwind, you would need to deploy the oars and start rowing.

Slight changes over the millennia occurred to these sail plans, but they still remained largely the same, with five main types of sails suspended and controlled from various rig iterations, up until the 1800s. The five main types of sails were:

1. Square sail hanging from a yard athwartship

2. Square sail hanging from a yard fore-aft

3. Square sail hanging from a yard fore-aft with another yard at the bottom

4. Triangular sail hanging from a yard fore-aft

5. Triangular sail hanging from a yard fore-aft with another yard at the bottom

As you can see, there wasn't much variation in these five sails. They are pretty much a square sail or a triangular sail hanging from a yard, with either nothing at the bottom or another yard at the bottom. Now that we have the main types of sails outlined, we can look at how they have been set on various rigs throughout the ages.

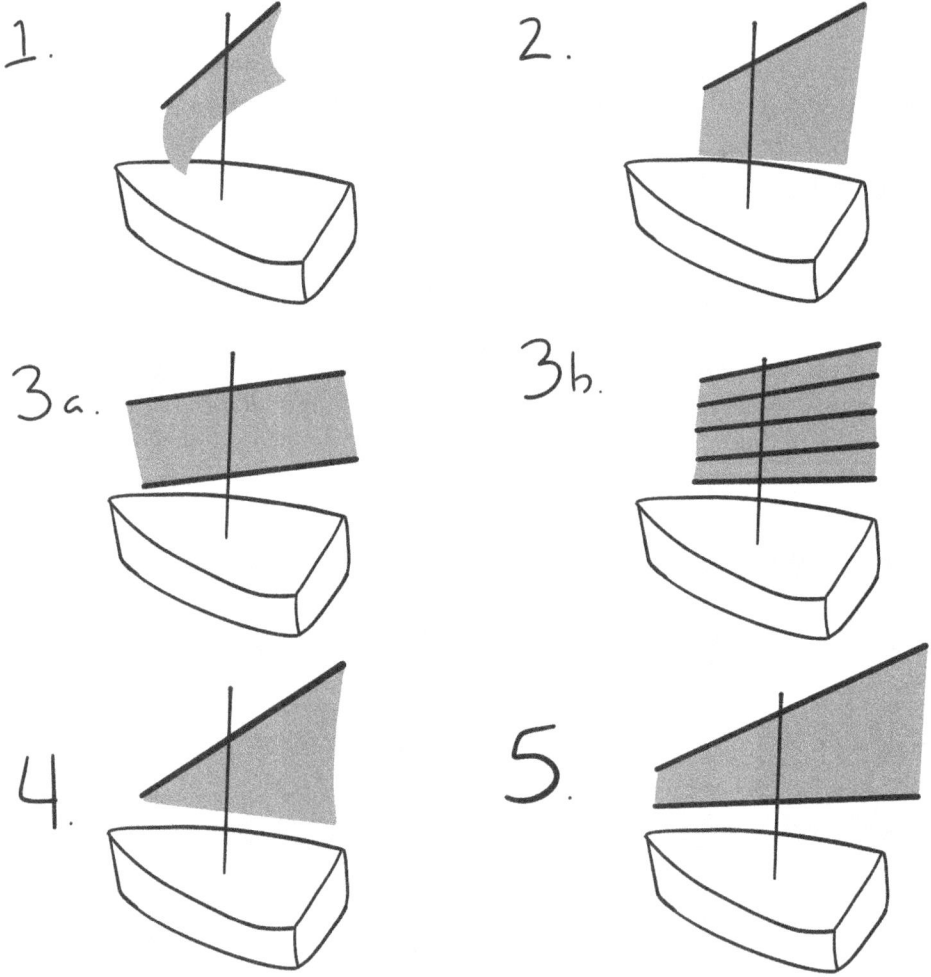

In this list, I didn't specifically mention junk sails because they are a repeating pattern of the third sail—a series of square sails set fore-aft with a batten in place of the yard in the middle sections.

AGE OF DISCOVERY

Starting in the 1400s, ocean exploration and trade were carried out by large, multi-masted, square-rigged ships. This is when the major moments in sailing history were occurring, such as Columbus sailing to the "West Indies" and the Portuguese making trade routes around Africa to China and India, as well as across the Atlantic to Brazil.

In my personal opinion, this is when ships became much more interesting, as the rigs were more thought-out and set up for an intended purpose. The importance of the keel in allowing the boat to sail to windward was discovered and, as a result, more effort was put into developing the rig to support more sails, whereas before, sails were only for downwind and oars were for upwind passages.

Some notable types of sailing ships that came from this era are:

- **1300s CE:** The **carrack**, a sailing ship instrumental in the Age of Discovery. This is the evolution of the cog, with three masts, the foremast and mainmast fitted with square sails, and the mizzen mast fitted with a lanteen sail.
- **1400s CE:** The **caravel**, a small maneuverable sailboat used by the Portuguese to explore the coast of West Africa.
- **1500s CE:** The **xebec**, a fast, coastal sailing vessel used in the Mediterranean Sea.
- **1500s CE:** The **galleon**, a massive ship with three decks and three masts, the front two sporting square sails and the aft mast rigged with a lateen sail. Galleons were instrumental in sixteenth-century naval battles and protecting maritime superiority in various territories.

Up until the 1600s, the sailboats were all named based on their construction methods and hull styles, with little emphasis on the rigging. For example, Columbus had *La Niña* rerigged from a lateen rig to a square rig in the Canary Islands to improve her sailing ability in the ocean, but the change in sails and rigging did not change the fact that she was still referred to as a carrack.

GOLDEN AGE OF SAILING

The golden age of sailing started in the 1450s and lasted until the 1800s. The ability to sail upwind was discovered, unlocking the entire upwind world to exploration and trade. A massive race to develop a better boat was unleashed, as this would bring countless riches to the country that could get there first!

Finding new routes and getting there faster would secure wealth, which spurred constant innovation in new sail plans and the rigs to support them. For the first time, sailboats were named based on their rigs instead of their hull styles, and this is the

THE RIGGING HANDBOOK

time period we will be focusing on as today's modern rigs were actually born during this time period.

The most important rigs from this era are:

Brig: A brig has two masts with square-rigged sails. The design became popular in the mid-1700s, and they were used for smaller and shorter trade voyages near and around England.

Brigantine: Just like the brig, the brigantine has two masts, but the foremast has square-rigged sails while the main mast has fore-aft sails. The mainmast could have square-rigged sails up high, but the main distinction between the brig and the brigantine is the fore-aft mainsail. Brigantines started in the late 1700s and were generally smaller than a brig, but were also one of the fastest rigs at the time. This made them especially popular among pirates in the British colonies before 1775.

Barque: A barque, barc, or bark is a vessel with three or more masts in which the foremast(s) and mainmast are square rigged and the mizzen is rigged fore-aft. In the mid-1700s, barques were a very popular trading vessel as they could match the passage distances of fully rigged ships but with a much smaller crew, which meant operating the vessel was cheaper. They were superior at downwind sailing when compared to schooners but, thanks to the fore-aft mizzen sail, were better to windward

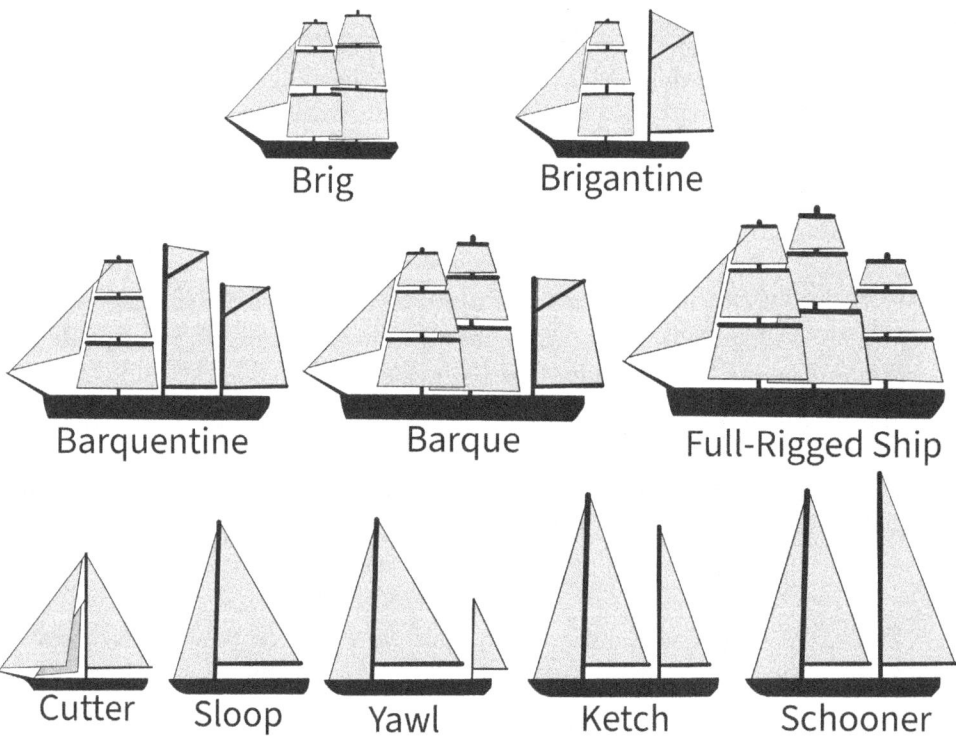

than a fully rigged ship. Barques were a compromise between the best elements of a square rig and a fore-aft rig.

Barquentine: A barquetine's rig is the opposite of a barque. Whereas a barque has fore-aft sails on the last mast and everything else is square rigged, a barquentine has the first mast rigged with square sails and everything else rigged fore-aft.

Clipper: Clipper ships could be any rig, as long as they were fast! The name comes from the word "clip," as in to move quickly. They were used as merchant vessels because they could get the cargo from origin to destination in the shortest amount of time possible.

Cutter: These were workboats where the people needed to get out to a destination in any weather. The most famous of cutters were the pilot cutters, which were used around England to get the harbor pilot out to the incoming cargo ship. The first one to the cargo ship got paid for the job, so pilot cutters had to be fast and seaworthy, as the waters around England are notoriously feisty.

Cutters have a single mast set at or aft of the 40% mark on the boat. This arbitrary point is found by dividing the length of the boat before the mast by the length of the boat. To be a cutter, the mast had to be at or behind the 40% mark, meaning that if you have a 40-foot boat, the mast had to be located at least 16 feet behind the bow. The mast could be farther back than this point, but if the mast was any forward of this point, say, 15 feet from the bow, it could no longer be considered a cutter. Having the mast farther aft allows for more space up front for a plethora of headsails, which has led to the more modern definition of a cutter (which will be discussed later in this chapter).

Sloop: Sloops, like cutters, have only one mast, but the mast is set forward of the 40% mark. Since the mast is set farther forward, the fore-triangle is smaller and, therefore, they usually only have one headsail.

Fully Rigged Ship: A fully rigged ship is a sailing vessel with three or more masts, just like the barque and barquentine, but on a fully rigged ship all the masts are square rigged. These were massive sailing ships with an extensive sail inventory. The square sails were stacked on each mast in a staggering order.

The sails were named for the mast they were on, so in this example, we will just look at the sails on the main mast. The sails proceeded from lowest to highest: mainsail, lower topsail, upper topsail, lower top gallant sail, upper top gallant sail, royal sail, sky sail, and moonraker. That is correct—each mast could have up to seven square sails stacked on the mast! If winds were light, they would add additional studding sails mounted to the sides of each sail to give the vessel speed in the lightest of airs.

Fully rigged ships were the ultimate tradewind cargo vessels, but they were only good on a downwind course. Lacking fore-aft working sails meant they struggled to windward, which is why they fell out of favor when steamships started to gain traction.

Fully rigged ships did have one fore-aft sail mounted at the lowest portion of the aft-most mast, and this sail was called the spanker. It was useful in helping the boat make it through a tack, as the vessel would easily get caught in irons and struggle tacking to windward.

Schooner: A schooner is a sailing vessel with at least two masts, where the aft-most mast is the tallest or equal in height to all the other masts, and referred to as the mainmast. The sails are all rigged fore-aft, but there is a variant called the topsail schooner that has a square-rigged topsail on the foremast and all the other sails rigged fore-aft.

MODERN RIGS

Now that we have progressed through about five thousand years of sailing rigs, it's time to discuss the modern rigs that you will encounter on contemporary sailboats.

ONE MAST
Sloop
The sloop is by far the simplest rig and therefore the most common rig you will encounter on the water. The mast is set farther forward of the 40% mark and the fore-triangle is small. To negate the issues of a small headsail, the headsail simply overlaps the mast and the clew comes rather far aft. The distance from the tack to the mast is deemed 100% and the distance that the foot of the headsail exceeds this measurement is represented by a percentage number that is greater than 100%. For example, if the fore-triangle length is 10 feet, and the headsail foot is 20 feet long, the headsail would be denoted as a 200% genoa. If the headsail foot is 25 feet long, it would be a 250% genoa.

The overlapping headsail offers greater advantages for upwind sailing created by the slot effect (see chapter 6).

Cutter
The old definition of a cutter still holds true, though the mast sure does seem to be riding the 40% mark on most cutters. Having the mast set farther aft opens up the fore-triangle for more and larger headsails to be flown within its confines. While a 40-foot sloop with a 200% genoa might sound like it has a massive headsail, that same sail on an equal sized cutter might only be a 125% genoa. All of this extra space allows you to have a more versatile sail inventory rigged and ready to deploy at all times. The headsail can be a large sail for lighter airs, and the staysail set on the inner forestay can be a smaller jib that is useful in heavier wind conditions.

This versatility has made cutters a favorite for ocean sailing where the winds are consistently stronger and you need to be ready to quickly change the size of the sails you are flying; after all, there is no safe harbor to seek in the middle of the ocean.

Slutter

A modern-day blending of the sloop and the cutter is often seen on sloops that are set up for bluewater cruising. The owners of these vessels want the versatility of the various headsails at the disposal of a cutter, but they don't have the fore-triangle space to hold a sail set on an inner forestay. The solution to this is to have the sail start on the deck a few feet aft of the headsail, but run up to the masthead, just like the headsail does. This allows you to run a larger genoa on the headstay and a smaller jib on the slutterstay.

Solent

The solent rig is a variation of a sloop where there are two headstays set fore-aft of each other. While this sounds a lot like the slutter setup, they are set so close together that you can't fly both headsails at the same time. This gives you the versatility of having two different cuts and weights of headsails at your disposal and completely deployable from the cockpit, just not at the same time.

TWO MASTS

Schooner

The schooner is considered by most sailors to be the most beautiful rig in existence. It looks incredibly pleasing to the eye, as the masts and sails also grow in size as you move farther aft. The defining quality of a schooner is that it must have at least two masts and the aft mast must be taller or equal in height to the forward mast.

The forward mast is the foremast, and the aft mast is the mainmast. The mainmast usually has a mainsail set on a boom, but the foremast doesn't always have a sail set on a boom. There are many variations of schooner rigs, each named for the special traits that make it slightly different from the normal schooner.

The most common variation is called a staysail schooner, where the foresail area is occupied by the staysail of the mainmast. Filling in the top portion of this open triangle will be another sail called a fisherman. If the wind is light, this vast span between the masts can be occupied by the best-named sail in all of sailing: the gollywobbler!

Sadly, there are no production boats being made with schooner rigs, so if you are lucky enough to come upon one of the few custom-built schooners of the world, you will be the envy of us all.

Ketch

A ketch is traditionally a two-masted sailboat where the aft mast is smaller than the forward mast and set before the rudder post. The forward mast is called the mainmast and the aft mast is the mizzen.

While sailing, if you see a two-masted sailboat on the horizon, you will not be able to see the rudder post to determine what kind of sailboat you are looking at, but if the mizzen is pretty big, then it is probably a ketch.

Ketches are numerically denoted in percentages, in relation to the size of the mizzen to the main. An 80% ketch will have a mizzen that is 80% the height of the mainmast, while a 40% ketch will have a mizzen that is only 40% the height of the mainmast. The larger the mizzen, the more it helps to drive the boat forward through the wind and waves. There is no 100% mizzen on a ketch as this would imply that the two masts are the same height and the ketch would become a schooner.

Ketches were considered to be the ultimate cruising boat as they offer great versatility in the sails that can be set for various wind conditions. If the winds are light, you can set all the sails, but as the wind builds, you can reduce the sail area incrementally.

The sail area is also spread out over a greater length. On a sloop with 1,000 square feet of sail area, you are going to have only two huge sails. The mainsail will be around 500 square feet and the genoa will also be around the same. This means that when you need to sheet or trim the sail, you have to pull 500 square feet of sail with a single line. As you can imagine, this is a very difficult task and the loads on the gear are monumental. That same 1,000 square feet of sail area could be broken up into three sails minimum, as some ketches have two headsails to further break up the sail area. If you take that same sail area and break it up over an 80% Ketch, you would be looking at about 330 square feet for the headsail, 420 square feet for the mainsail, and 260 square feet for the mizzen sail.

You might see a two-headsail setup on a ketch and think "ah yes, the cutter-rigged ketch," but this is a misnomer as a cutter only has one mast and a ketch has two masts. If the ketch has more than one headsail, it is still a ketch the same way that a fully rigged ship is one regardless of how many sails are stacked on each mast.

This practice of breaking up the sail area not only makes each sail easier to manage, but also shortens the sails, making them have a lesser tendency to heel the boat over as the lever arm is shorter. This means that the boat will also sail more upright, which translates into greater comfort for everyone on board.

Yawl

On a ketch, the mizzen is forward of the rudder post, while on a yawl, the mizzen is aft of the rudder post. Just like before, you can't see the rudder when the boat is sailing on the horizon, but a classic sign that it is a yawl and not a ketch is the mizzenmast on a yawl tends to be tiny and the mizzen boom hangs off the back of the boat.

The mizzen on a yawl is too small to provide any actual drive through the waves, and the additional rigging creates added windage, which reduces upwind performance. All of this sounds like the yawl rig holds no advantages, but this is not the case.

The tiny mizzen on a yawl works like an air rudder to help steer and guide the vessel. The mizzen sail is used more as a balancing sail, to alleviate the work of the rudder in the water and actually help steer the boat. Before the time of reliable autopilots, the yawl rig was favored for long ocean passages for this very reason. They also boomed in popularity during the Cruising Club of America (CCA) racing rules era (the 1930s–1960s), when a mizzen headsail on a yawl was not counted towards the boat's sail area; thus, any sail set on the front of the mizzen was simply free sail area!

On long ocean races where you will be on the same tack for weeks, you could easily set a mizzen staysail and zip along quicker thanks to the additional sail area. Yawls gained popularity as a "Rule Beater" racing sailboat, and quickly vanished when the rules changed from CCA to International Offshore Rule (IOR).

CHAPTER FOUR
The Mast Itself

Where a mast is placed and how tall a mast stands relative to other masts plays the defining role in determining the name and characteristics of a rig, but what about the mast itself? How are masts classified and categorized, and how do these differences affect sailing performance?

The main classifications of the mast are determined by the material it is made of, followed by where the mast heel rests. After that, masts are distinguished by the number of spreaders they have and where the headstay attaches. The next consideration is the shape of the mast section; and lastly, is the mast fixed or movable. All of these factors play a major role in determining the sailing characteristics that the mast will provide.

Before going into detail on the mast itself, we need to look at a few things related to weight and the effect of its position. The more something weighs, the more effort is needed to make it start moving. When you flick a ping pong ball, it will shoot across the room, but the same flick will not have the same result with a bowling ball. The greater the mass an object has, the greater its inertia and the greater the force it will require to change its motion. This means that a heavier boat will require more force to move it, which is why racing sailboats are made out of the lightest materials possible.

While the weight of the boat is important, more important is the *placement* of the weight. I'm sure you have heard the term "pound-foot" before. The basic definition is the amount of torque required to support 1 pound, 1 foot away from the point of rotation.* For those with an affinity for the metric system, this torquing force is represented by Newton-meters, where the force is 1 Newton supported 1 meter away from the point of rotation. Since Newtons are force and not weight, it is easier to just use pound-foot for the comprehension of this system. Picturing a static weight is easier than picturing a weight being accelerated a full meter per second every second, which is what a Newton is.

---

* *Pound-foot* is the measure of torque, while *foot-pound* is the measure of work or energy to move 1 pound a distance of 1 foot. They are often used interchangeably because "foot-pounds" sounds better than "pound-feet," but this is incorrect.

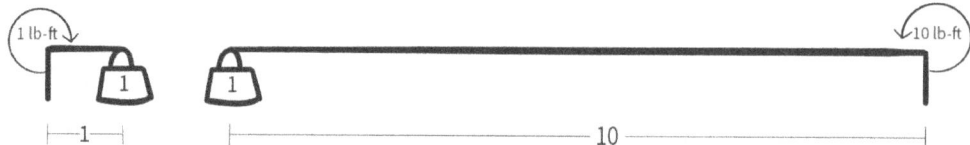

If you have a 1-pound weight at the end of a 1-foot pole, the force required to hold the pole parallel to the ground would be 1 pound-foot. If you extend the pole and hold the same 1-pound weight at the end of a 10-foot pole, suddenly the same weight feels a whole lot heavier! You need to exert 10 pound-feet of torque to hold that same 1-pound weight parallel to the floor. The weight didn't increase, but the force you need to supply to hold it there did.

On a sailboat, the two parts of the boat that can have the biggest impact on the uprightness of the boat are the two longest parts extending from the hull: the keel and the mast. The keel extends downwards and counteracts the mast that extends upwards. If you have a 10-pound weight at the top of a 50-foot mast, you would have 500 pound-feet of torque trying to pull the boat over onto its side. To resist this, you could have a 100-pound weight at the bottom of a 5-foot keel, and suddenly everything is balanced! Now you have two pendulums extending from the boat: One goes up and one goes down. Longer and heavier pendulums will slow the rolling motion of the boat, as a heavier and longer pendulum will swing more slowly; this might sound like a good thing, but it is not the only force that will be acting upon the sailboat. The sails will exert the greatest force and cause the boat to heel over to one side. The weight in the keel is there to counteract this force as well. If the first 100 pounds in this example were needed to counteract the weight aloft of the mast, then we will need to carry even more ballast for the keel to do its primary objective of resisting the heeling forces caused by the sails. All of this added weight means that the whole boat ends up being heavier and requires more wind energy to sail.

Weight is a major concern on a sailboat, and now you know why weight aloft is the worst kind of weight. This is why oversizing your rigging is actually detrimental to the performance of your boat. (We'll discuss this further in chapter 13.) By reducing weight aloft, the keel can put more of its weight behind resisting the forces acting upon the sails, which means you can carry more sail in the same kind of wind, thereby directly improving the sailing performance of the boat.

The best way to control the weight of the mast is to choose the right material. Each has its own benefits and drawbacks, which all need to be considered while also looking at their respective weights.

MAST MATERIAL

The material of the mast has a lot of attributes that play into the rigging needed. Some materials are inherently stronger than others and can support the loads on their own while other materials might need more help supporting the loads in the form of standing rigging. Temperature plays an important role with rigging as well, as different materials expand and contract at different rates as they cool or warm up. This change in size might be rather small, but when extrapolated over the long length of a spar, the change can add up to be a significant change in tension. The three main materials for spars are wood, aluminum, and carbon fiber.

Wood

Wood has been the spar material of choice for thousands of years. The first masts were nothing more than a hewn tree that was selected due to its straightness and length. As mast heights grew taller, it soon became impossible for a single tree to make up the entire spar, and advanced woodworking techniques were employed to connect various pieces of wood to make taller and longer spars for the purpose.

Wood masts can be made solid, for added strength, but this comes at a significant weight penalty. The ideal wood of choice for spars is clear grain Sitka spruce, which only grows in Alaska. Clear grain simply means that there are no knots in the wood that would detract from the structural properties of the wood fibers that compose the spar. Sitka spruce weighs 29.2 pounds per cubic foot, which means that if you have a solid square box-shaped mast that is 12 inches by 12 inches, every foot of height will weigh 29.2 pounds. If your mast is 50 feet tall, the mast will weigh 1,460 pounds. That is a lot of weight and it is also placed high in the sky, which will negatively affect the performance of the boat.

The goal of any sailboat is to have the weight lower than the center of gravity and center of rotation to resist the force of the wind on the sails. This is why all the weight is placed at the bottom of the keel where it has the greatest effect on the performance of the boat, not at the top of the mast! As you move farther from the hull, which is the center of rotation on a sailboat, every pound you place is multiplied by the number of feet from the center of rotation. On a 50-foot mast, every pound at the top exerts 50 pound-feet of torque on the boat, while every pound at the bottom of a 10 foot keel exerts 10 pound-feet of torque to resist the forces acting on the mast.

Having a 1,460-pound mast is far from ideal. The weight of the top 1-foot section of the mast would weigh 29.2 pounds and suspended at the end of the 50-foot pole, would result in a torque equal to 1,460 pound-feet when the boat is heeled over. Imagine the weight in the keel to oppose this kind of weight aloft, not including the other 49 feet of heavy mast below this part. The best way to reduce the weight aloft in this situation is to hollow out the mast. Air weighs less than wood, so replacing the wood with air in all the places that don't need the strength is a great way to save weight aloft without sacrificing strength.

If the mast were made out of 4 boards that were only 2 inches thick, suddenly, each 1-foot section of mast would weigh 15.4 pounds. The same 50-foot mast would weigh 770 pounds. This is a weight savings of over 47% and, more importantly, the last 1-foot length of mast would now be exerting only 770 pound-feet of torque instead of 1,460 pound-feet!

As the mast gets taller, each 1-foot section of the mast exerts a proportionally greater torquing force than the last. In the simplest form, you can think of it as the first foot exerts just its weight in pound-feet, while the second exerts double, and the third foot exerts triple.

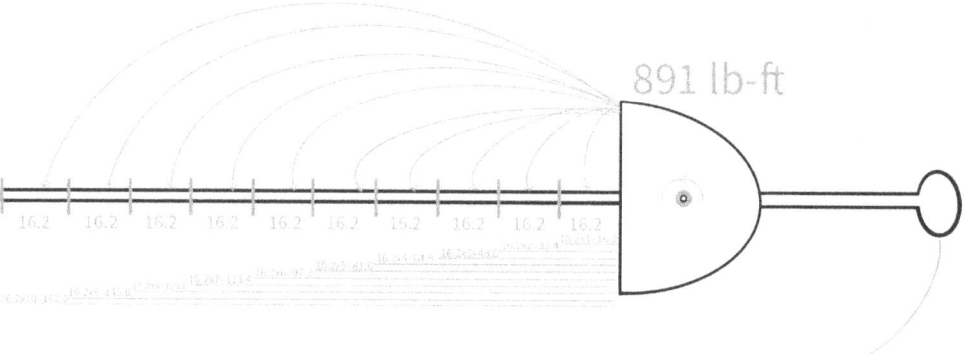

The taller the mast, the greater the impact the weight has on the balance of the whole system. This makes a longer lever arm that is subjected to even greater torquing forces when you think about the actual sail filling with wind that far up in the sky! To rudimentarily calculate the total torquing force exerted by the weight of the mast itself, you will need to do a bit of math.

To make the math easier, we will simply take the mast and divide it up into 1-foot sections, accounting for the weight of each section, we can then roughly calculate the total number of pound-feet that the spar itself will exert. For this example, we will use the weights of a hollow mast section at 16.2 pounds per foot of mast.

Length	1	2	3	4	5	6	7	8	9	10
Weight of section	16.2	16.2	16.2	16.2	16.2	16.2	16.2	16.2	16.2	16.2
Pound-feet of section	16.2	32.4	48.6	64.8	81.0	97.2	113.4	129.6	145.8	162.0
Pound-feet added up	16.2	48.6	97.2	162.0	243.0	340.2	453.6	583.2	729.0	891.0

THE MAST ITSELF

As you can see, the total torque in pound-feet of the mast increases dramatically as the mast gets taller. The math is also simple enough to do for a small distance, but if you have a 50- or 70-foot mast, the math becomes a bit overwhelming. This is where your old nemesis from high school comes back to haunt you: calculus!

$$\sum_{n=1}^{length\ of\ mast\ in\ feet} (weight\ of\ each\ 1\ foot\ section)n$$

For a 10-foot mast where each 1-foot section weighs 16.2 pounds, the formula would be: $\sum_{n=1}^{10} 16.2n$; the result would be 891 pound-feet, the same that we got from our table. Now that we know the math works, we can see what a 50-foot mast would act like.

$$\sum_{n=1}^{50} 16.2n = 20{,}655\ pound\text{-}feet$$

A 70-foot-tall mast would be:

$$\sum_{n=1}^{70} 19.5n = 40{,}257\ pound\text{-}feet$$

Reducing weight aloft is critical to improving the performance of the boat as less ballast will be needed just to counteract the weight of the mast itself and that allows the whole boat to be lighter in weight, which will sail faster as a result.

If you are an adept in the world of mathematics, then you know that you can play with these values to improve the accuracy; $n = 1$ refers to n being 1-foot segments and the number at the top of Sigma is the number of sections. If you wanted to make n inches, you would simply need to multiply the number above Sigma by 12, and then divide the coefficient of the equation by 12. The result will be more accurate but about the same, so it's not really necessary to get that picky since we are talking in pounds on a boat that is weighed in tons.

For those of you who skipped the calculus refresher, there is a still easier way to come up with the total force exerted by the mast. Just use this simple formula instead:

(Mast Length ÷ 2) + 0.5 = Mast section to work with

Mast Section to work with × Weight of 1 foot mast section

= Pound-Feet of that Section

Pound-Feet of that Section × Length of the mast = Total Pound-Feet of the Mast Weight Aloft

For example:

$$(10 \div 2) + 0.5 = 5.5$$

$$5.5 \times 16.2 = 89.1$$

$$89.1 \times 10 = 891 \text{ pound-feet}$$

This is the same result as the complicated way of $\sum_{n=1}^{10} 16.2n = 891$ pound-feet.

Wooden masts tend to be on the heavier side of things, but they do have many advantages that other masts lack. First of all, they are made out of a material that can be worked on with basic tools in any part of the world. From the richest to the poorest ports, wood is a commonly used material and sourcing skilled material or labor is always possible. This is an excellent quality for sailors who plan on sailing to remote locations where the other mast materials might pose a bit of difficulty in this department.

Wood also is easy to inspect for rot, as long as it is varnished and not painted, as it will show up as black streaks on the surface of the wood. When wood starts to rot, it will become soft and lose its strength, eventually leading to a failure of the spar. Repairing a wooden mast is rather straightforward, as the infected area is merely cut out and fresh wood is scarfed in to replace the decayed wood.

Wood also has an advantage that is unparalleled by all other materials, it is beautiful! Why else would massive luxury yachts with high-tech carbon fiber masts have intricate paint schemes applied to the mast to make it look like wood? This beauty does come at a hefty price, and that is paid in sweat equity to maintain the spar in Bristol condition.

Since a wooden mast is by its very nature made of wood, it is subject to the types of changes that wood will go through when exposed to the elements. Temperature and humidity will play a role in the size of your mast. Wood naturally swells when it gets wet and shrinks when it dries. These changes can be mitigated by sealing the mast inside a shield of varnish, keeping the wood dry and relatively steady in size. The thin layer of varnish that coats the mast is very delicate, and the slapping of halyards, sails, blocks, and any other airborne debris will chip and nick the surface. If you have a varnished wooden mast, you will need to make, at a minimum, monthly trips up to the top with a can of varnish and a brush to touch up the coating and maintain it in perfect condition. If you slack off on these duties, the breaks in the varnish will allow water to intrude and begin affecting the wood. These will become the future spots of wood rot and will be much more costly and time consuming to repair than frequent varnishing trips up the mast.

THE MAST ITSELF

> **Pro tip:** Go all the way to the top of the mast without varnishing and then varnish from the top down. This will keep you and your clothes clean, as you won't touch any wet varnish—it will all be above you. When you get to the bottom, the mast will be freshly varnished and ready for another month of service.

The other factor that is harder to control is temperature. As the temperature changes, wood will change in size at a rate of roughly (3.5 × 10^-6) per Kelvin per meter of wood. This is known as a positive coefficient of linear thermal expansion, which means that it will expand as it warms and shrink as it cools. This is a tiny change, but spars are long and the effects compound! On a 62-foot mast (19 meters), the effects would be $\left(\frac{3.5 \times 10^{-6}}{K}\right)(19m)(40K) = 0.00266m = 2.66mm$ (0.10 *inches*) longer as the temperature changes from frigid to really hot. A change this small will go unnoticed to the casual observer, but your stays that are made out of a different material with their own coefficient of thermal expansion will notice the change.

The last element that plays a major role in the way your mast moves is the grain structure of the wood that was used to make the mast. This is solely in the hands of the carpenter who selected specific boards and arranged them in a certain way to make the mast. How the wood was cut and what part of the tree the board came from will have an effect on how much the wood will move as temperature and humidity fluctuate. This is where a knowledgeable woodworker makes a difference; the choices they made at the time of construction will have the biggest impact on how the mast will behave for the rest of its life.

Aluminum

Wooden spars used to be the standard, but as aluminum became cheaper to use, it quickly replaced wood as the material of choice. It is lighter and stronger than wood, which has made it the favorite and also the most common type of mast material you will see in any marina.

Aluminum is a "catch-all" for the actual material. Most sailboat masts are made out of 6xxx grade aluminum, varying from manufacturer to manufacturer; some will use 6061 while others will use 6082. This metal is very strong and hard, making it an excellent mast material thanks to its compressive strength. By comparison, aluminum weighs around 169 pounds per cubic foot, which is way more than the weight of wood, but since it's so much stronger, it can also be made much thinner. This makes for a much lighter mast given the same size.

For a 45-foot sailboat, the wooden mast can measure up to around 13.5 × 6.5 inches and be made out of 2-inch-thick planks on the front and back and 1.5-inch-thick planks on the side. For a similarly sized (strength-wise) 45-foot sailboat, the aluminum mast would be 9.4 × 6.1 with wall thicknesses of just a few millimeters and weigh in at just under 300 pounds for the whole spar. A mast this size is made of very thin aluminum and weighs about 5.6 pounds per foot. This is less than half the weight of the lightest comparable wooden mast!

Saving weight in the mast is the most important place to shave weight off, as the weight penalty is monumental when you compare the torquing the mast exerts on the boat when you start to heel over.

A 50-foot aluminum mast that weighs roughly 275 pounds total would only exert $\sum_{1}^{50} 5.6n = 7{,}140$ pound-feet of torque on the boat. This is a drastic decrease from the torque generated by any wooden mast, as the aluminum mast can be made so much thinner and lighter than a similarly sized wooden mast.

The weight savings of the mast mean that with less ballast or a shallower keel, a sailboat can now carry even more sail and therefore sail even better. This leap in sailing ability is why wooden masts were almost completely replaced by aluminum masts decades ago.

While aluminum is incredibly strong and much lighter than a comparable wooden mast, it does have a few detracting qualities. First, aluminum is subject to corrosion and will quickly turn into a white powder, known as aluminum oxide. Keeping the aluminum sealed and protected from the elements is an important step to preserve the structural integrity of the metal. Aluminum also suffers from galvanic corrosion, which is caused by dissimilar metals being in contact in an electrolyte solution. Sadly, saltwater is all around a boat and qualifies at the necessary electrolyte solution to jumpstart the galvanic corrosion. The most common dissimilar metal introduced to an aluminum mast is stainless steel, which is used in all the fasteners on the aluminum spar. Using an isolator like lanolin or Tef-Gel will negate this issue, but it is important to check every screw in the mast for the classic appearance of bubbles under the paint around the fasteners.

While humidity has no apparent effect on the size of an aluminum mast, temperature changes will have a drastic effect on the length of the spar. The coefficient of thermal expansion of aluminum is (23.1×10^{-6}) per K. If we look at a 62-foot mast (19m), we will see a change of 17.56mm over a 40°C change.

$$\left(\frac{23.1 \times 10^{-6}}{K}\right)(19m)(40K) = 0.017556m = 17.56mm \ (0.69 \ inches)$$

This is a considerable difference in size when extrapolated over the length of a mast, and this fluctuation can lead to some detrimental effects with different types of rigging. Stainless steel has a very similar coefficient of (16.0×10^{-6}) per K. While the aluminum mast grew by 17.56mm, in the same temperature swing, the stainless-steel stay would grow 12.16mm, so there would only be about a 5mm difference between the two materials, as they grow and contract with each other. Dyneema fibers have a negative coefficient of linear thermal expansion, which means they will change size in the opposite direction of an aluminum mast. Basically, as it gets cold, the aluminum mast will get shorter and so would a steel stay, but a synthetic stay would grow longer, exacerbating the effects of the change in temperature.

Carbon Fiber
While aluminum provided a massive leap forward in sailing performance from wooden masts, carbon fiber masts are the next radical leap forward in performance. A carbon fiber mast weighs about 40–60% less than a comparable aluminum mast. As you know, the reduction of weight aloft has the biggest role on the weight usage of the boat. Having a lighter mast means that the same weight in the keel will be used more for resisting the forces caused by the sails instead of just trying to counteract the weight of the mast as it hangs off to the side.

Since the spar is made up of smaller strands that are woven together intentionally, it is possible for the first time since wooden masts to put more material, and therefore more strength, exactly where you need it to be. This means that areas of higher load can be thicker and stronger while areas that don't need to be as strong can be made thinner and lighter. The result is a spar that weighs even less than if it were a uniformed thickness, as it is only as strong as it needs to be where it needs to be.

Carbon fiber also barely changes in size as temperatures change. The coefficient of thermal expansion for carbon fiber is (-0.8 x 10^-6) per K.

$$\left(\frac{-0.8 \times 10^{-6}}{K}\right)(19m)(40K) = -0.000608m = -0.608mm \ (-0.024 \ inches)$$

In other words, as the temperature swings from frigid cold to boiling hot, a carbon fiber mast holds the same size! It has a negative coefficient of linear thermal expansion but the coefficient is so small that it barely moves at all.

Carbon fiber is also wildly more expensive than any other type of mast material. These costs may seem prohibitive for someone in a pocket cruiser, where the mast might cost more than they paid for the entire sailboat, but on super yachts, the cost of the carbon spar seems less dramatic. A $100,000 mast on a $20,000 sailboat is ridiculous, but on a $1.2 million sailboat, it seems like a drop in the bucket. These larger sailboats also benefit from the reduced weight aloft, as the masts on these boats will be approaching, if not exceeding, hundreds of feet in height. The leverage that the weight of the mast exerts on the boat is multiplied dramatically and the cost of additional ballast would start to outweigh the benefits of employing a less expensive spar.

MAST STEP LOCATION

A major distinction between types of sailboats is where the mast heel is situated on the boat. The bottom of the mast is called the heel and it stands on the mast step. The location of the mast step on the boat plays a major role in the characteristics that the mast will have. The two configurations are deck stepped and keel stepped.

At first glance, a deck-stepped mast has the mast step on the deck, whereas a keel-stepped mast has the mast coming through a hole in the deck and the mast coming down all the way to the keel. Basically, a deck-stepped boat will have an open salon with no mast coming through, while a keel-stepped boat will have the mast barging into the middle of the boat, normally coming into the salon, but sometimes into the head or other quarters. While the interior aesthetics are the most apparent, the mere location of the mast step plays a huge role in determining the sailing characteristics that you will get out of your mast and rig.

In general, a deck-stepped mast will be bendier and a keel-stepped mast will be stiffer. This means that racing sailboats like to have a deck-stepped mast, as this allows for greater tuning and tweaking of the rig, and cruising sailboats like to have a keel-stepped mast, as these sailors "want something that works" and don't want to fiddle with it too much.

The bendiness of a mast is the result of a few factors: length overall, length of each panel, and the thickness of the walls. Some of these make sense intuitively, while other features may take a little experiment to wrap your head around. For this next section, I recommend getting out a package of uncooked spaghetti noodles, as these will be your "masts" during the experiments.

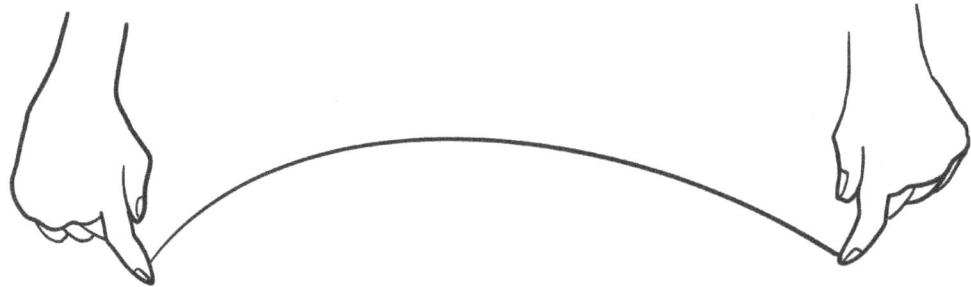

The first feature to look at is the length overall. The longer the noodle, the easier it bends. If you take a very short noodle and try to compress it between your fingers, it will remain very stiff and won't bend much at all. A much longer noodle will bend with ease and will also bend much further than the short noodle could.

The same holds true for masts. If you have two identical masts, made out of the same cross-sectional size and shape, the longer mast will be bendier. The first and most logical method used to stiffen the mast is to make it bigger and thicker. The obvious downsides of this practice are that the mast will also become heavier, and weight aloft is your biggest enemy.

As you can imagine, a thicker spaghetti noodle would naturally be stiffer and more resistant to bending than a thinner spaghetti noodle of the same length. If you happen to have some angel hair pasta on hand, you can compare the bendiness of that to the thicker spaghetti noodle. If you start bending the noodle to its limits, the noodle will crack and snap, just like your mast would if it were unsupported and bent to this extreme.

We have established through this test that longer noodles are bendier than shorter noodles. We also know that sailboats need tall rigs to hold bigger sails in the sky. What if we stack shorter sections of noodles together to make the rig taller incrementally? Well, that is precisely what they did up until just a few hundred years ago before metal masts came to dominate the sailing world.

This method worked well for the time, as the limiting factor was twofold: weight aloft and finding trees that were tall enough. Making the top masts thinner and lighter meant that they would require less ballast to counteract their weight aloft. Once metal masts came onto the ocean, wooden spars were quickly replaced, as these spars could be made into limitless lengths, and at a drastic weight savings.

Instead of stacking various shorter masts together to create a taller rig, we can simply take a single length of a metal mast and support it along the way up. Each span of mast between supported areas is referred to as a panel, and the length of the panel is also called the unsupported length.

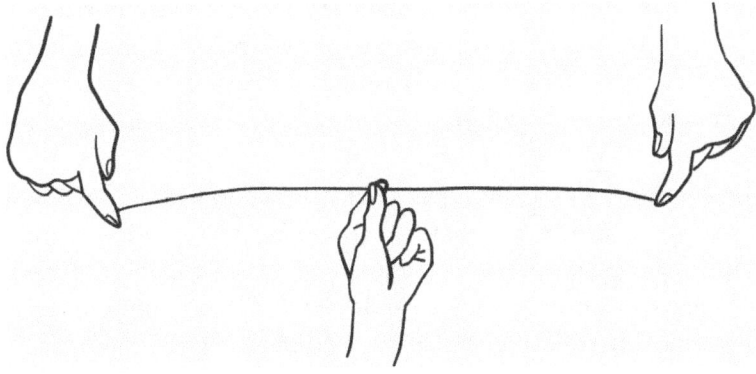

Grab a friend or an uninterested bystander and have them help you out for this next experiment. Hold the uncooked spaghetti noodle with your fingertips and compress the long unsupported noodle. With almost no effort at all, you are able to bend that noodle into a nice arc! Now straighten the noodle back out, have the other person pinch the noodle in the middle and give it another test. The noodle will be much stiffer and harder to bend with your fingers. Suddenly, the unsupported length of your noodle was cut in half and as a result, the size of the panels is smaller and the whole apparatus is stiffer. If they pinch the noodle in two places, the noodle will become even stiffer as the unsupported lengths are shorter yet.

The noodle is exactly the same weight, diameter, and length—the only difference is the amount of support that it has. This carries over to a sailboat directly! The more stout the mast is, the less support it needs from rigging. This means that a shorter and heftier mast will not need to have any rigging supporting it, also called an unsupported mast. However, it is also heavier aloft and will require more ballast to resist it, which makes the whole setup harder for the wind to push around. By contrast, if the desired mast is to be taller and/or lighter, it will need to have some external support to help it out.

Each point that the mast is attached to rigging acts the same as a pair of imaginary fingers pinching the spaghetti noodle. The unsupported lengths become shorter and the number of panels increases, which yields a stiffer mast for nearly the same weight. Naturally, there is a point of diminishing returns as the mast gets thinner but there is more weight in the rigging to support this ultra-lightweight mast. You can't have a mast the size of a human hair supported by a spider web of steel cables and claim that it will perform better than a setup with normal-looking rigging.

On a deck-stepped sailboat, you will have at a minimum two points of attachment: the mast step and the masthead. In between, you will have as many or as few additional points to divide up the unsupported length and give you the desired amount of mast stiffness.

On a keel-stepped sailboat, you will always have one additional point of attachment than on a deck-stepped sailboat, thanks to the mast-deck interface. On a keel-stepped sailboat, you will have at a minimum three points of attachment: the mast step, the mast-deck interface, and the masthead. In between the mast-deck interface and the masthead, you will again have as many or as few additional points to divide up the unsupported length as needed.

Why would you want a stiffer mast or a more noodly mast? Well it depends on what you want your boat to be able to do. The shape and position of the mast plays a pivotal role in the performance of the mainsail, which will be discussed further in chapter 6. Imagine if you are in the lead and your rival is right on your stern. You want to trim the mainsail to make it generate more power, and to do so you need to change the shape of the mast. You have a stiff keel-stepped mast with very basic standing rigging and your only tool at hand is to adjust the backstay. Your rival on his boat with a bendy deck-stepped mast is tuning his mast to perfection with his running backstays and checkstays he has at hand, and is quickly gaining on you. This is the same feeling as when you need a screwdriver but only have a hammer!

For the racer, a bendier mast that can be adjusted into limitless iterations of itself is ideal because they can tune it to perfection for each sail they are flying, the wind conditions, and the point of sail they are on. In general, you want the mast to be straighter and farther forward for downwind with looser rig tensions so that the sails will fill in more, while upwind sailing favors a tighter rig with a mast that is raked back and bent back as well.

The cruiser is not racing, and having all this extra rigging to deal with is rather annoying when friends have boats where they haven't touched the rigging in years—a "set it and forget it" kind of mentality. For a cruiser, having a stiffer mast that won't require as much tuning and adjusting is ideal. The sails will fill and the boat will move well enough to make everyone onboard happy about the journey ahead. For these sailors, having a keel-stepped mast is ideal because the mast derives some of its support from the mast-deck interface, making the mast stiffer without any added weight aloft.

SPREADERS

As the mast gets taller, it becomes impossible to have a deck wide enough to reach the masthead at a favorable angle to generate enough lateral pulling force. The minimum angle for a stay to approach a mast is 12.5°. If the angle is greater than 12.5°, then it will generate enough lateral pull to be effective. If the angle is less than 12.5°, then it will simply be pulling down and not really to the side, so it will actually be adding to the compressive forces on the mast without offering any benefit or support. In other words, it would just make things worse.

Instead of limiting our mast height based on deck width to achieve this minimum angle of 12.5°, we can simply do a little trickery to move the stay farther away from the mast on its way up. This is what spreaders do—they simply spread the shrouds out to the side so that when they make their final approach to the mast, they come at it at an angle greater than 12.5°.

As you know, every force has an equal and opposite reaction. When you push the shrouds outboard to create a better angle of attachment to the mast, you are also inducing a pushing force on the mast at the base of the spreader.

To make the force on the spreader purely compressive, the spreader needs to have the same angle made between the spreader and the shroud above and below. This is known as having the spreader bisect the stay, and results in the spreader having a gentle upsweep. If the spreader is purely horizontal, the compressive force would also be driving the spreader downward, and if the spreader tip is not properly seized, the spreader could easily fold down as everything comes crashing over. The same holds true if the spreader is too high up—the compressive force would cause the spreader to fold upwards. Having the angle made by the spreader and the stay the same above and below means that these forces cancel out and the resulting force is purely compressive in nature.

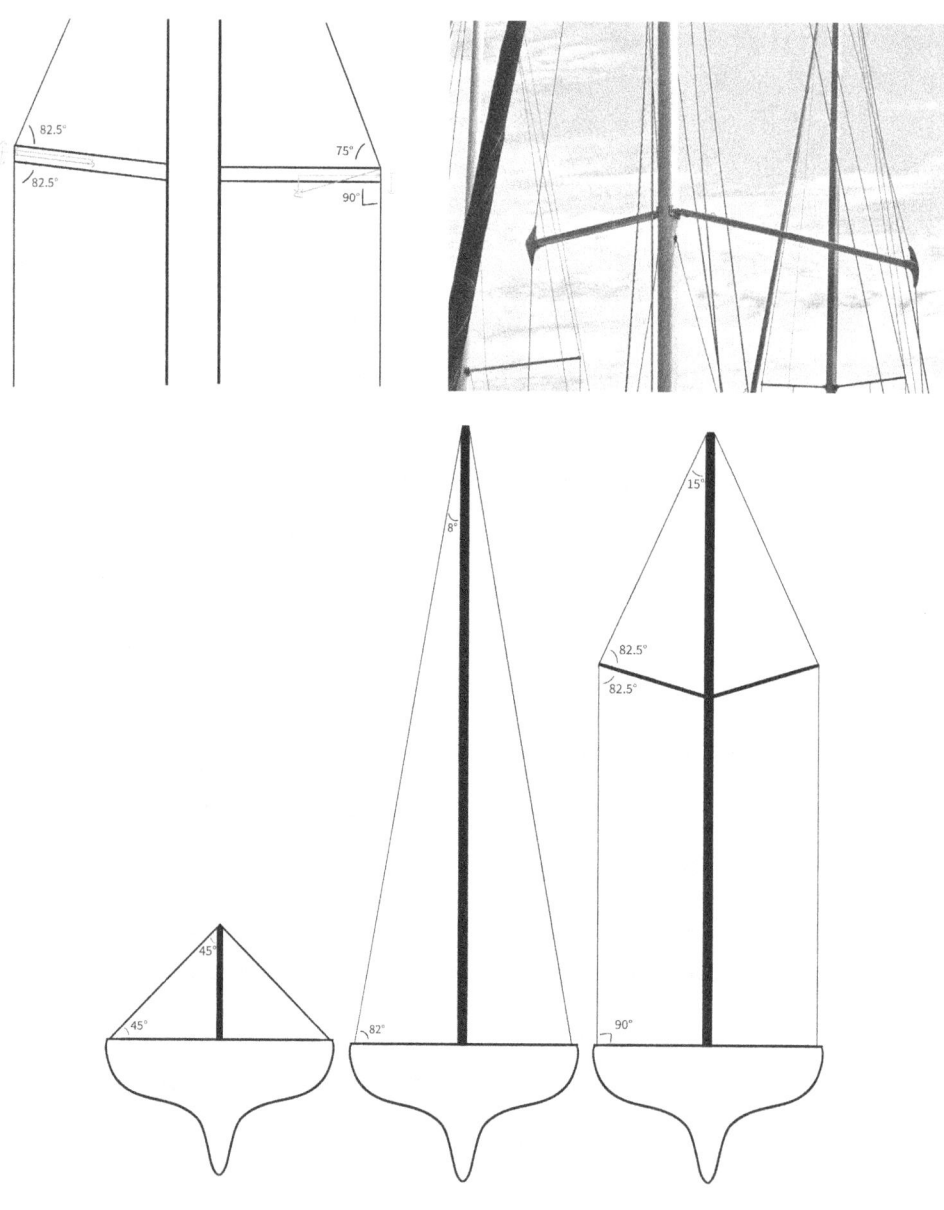

THE MAST ITSELF

If the forces from the spreaders were equal on both sides and simultaneous, then that would be the end of the book, but seeing how there are a lot of pages in your right hand, you know that is not the case. While both sides of the rig will experience the same kinds of loads, the forces are unilateral. The windward side will be doing all of the work while the leeward side relaxes. When you change tacks, the port and starboard designation of leeward and windward will swap and the other side will take the load while the prior working side relaxes.

When the sails fill and pull the boat over to leeward, the cap shroud becomes loaded and the tension on it increases dramatically. The stay runs through the tip of the spreader, giving it a non-direct path, which the tension it is experiencing promptly tries to rectify. This puts the spreader under a great compressive load as the spreader tip tries to be pushed to leeward. That force is transmitted straight down the spreader and onto the mast section where the spreader base attaches. This force, if unopposed, would cause the mast to push to leeward at this point, resulting in a nice bend in the mast.

As the spreader moves laterally to leeward, the cap shroud also moves to leeward, which diminishes the angle that it reaches the mast and also its effectiveness at doing the task at hand. To counteract this force, an additional stay is placed just below each spreader base. This stay pulls out to counteract the inward push caused by the spreader base.

On a single spreader rig, this stay is the lower stay, named so for supporting the lower portion of the mast. On a multi-spreader rig, the stays that span the distance between the lower spreader tip and the next spreader base are called intermediate stays because they are neither the lowest nor the highest stays; as such, they are the intermediate stays.

Having spreaders makes it very easy to identify the characteristics of a sailboat from a distance, as they are very visible and high up in the sky. If you see a boat with no spreaders, it will have a short and stout mast, as it needs no added support between the deck and masthead. This setup is common on gaff-rigged sailboats, where the mainsail is attached via hoops that run around the mast. Any spreaders or rigging on the mast would be in the way of the mainsail and dictate the maximum length of the sail's luff. On some gaffers, you will see a single spreader at the very top of the mast; this is to give the mast as much height as it can, while still not having anything along the run of the luff.

On sailboats from the past hundred years, you will normally see at least one spreader placed in the middle of the mast. This spreader divides the mast into two panels on a deck-stepped mast and three panels on a keel-stepped mast. These are common on older sailboats under 50 feet. These rigs are heavy, as they do not need much rigging to derive support. At the same time, they don't bend much, which is why they don't need much rigging to support them and also why they are not very good at tweaking to gain the most out of your sails. Having less rigging, though, means that the mast will generally hold its shape well enough on its own and will not require as much tuning or fiddling to get it perfect. It just is "good enough" and stays "good enough."

On contemporary sailboats, you will now normally see at least two sets of spreaders. The additional spreader is driven by two major factors. First, racing sailboats have more spreaders, and people want boats that look like racing sailboats. Second, the mast can be made out of a lighter material.

The reason racing sailboats have more spreaders is because this is a great way to reduce weight, and it also grants the racers ultimate control over the sail shape. For regular sailboats, the reduced weight aloft leads to better sailboat performance at a reduced weight overall. The keel can also be lighter and just as effective, and for the sailboat manufacturer, less ballast also lowers the cost to build. This makes a perfect situation where the manufacturer can save some money on materials and the consumer wants the result.

For your average sailor, they are not going to be tweaking and tuning the rig for every point of sail they change to, which is why they then develop issues with the rig being out of tune. Since the mast relies more heavily on the rigging to maintain the desired shape, maintaining the rigging is more imperative and also more complex.

THE MAST ITSELF

As the amount of rigging increases, so does the amount of tuning required, as does the complexity of the tuning to get it back to perfection. On any rig, you need to make sure that the top of the mast is directly over the middle of the boat and that the top of each panel is also in column with the rest of the mast. The more panels you have, the more panels you need to tune!

On a single spreader rig, this means that you will need to position the masthead over the boat and then the middle section where the lowers connect. On a two spreader rig, you need to do the same but also need to adjust the intermediates to get the second spreader section in column with the rest of the mast.

On a single spreader rig, the common problem of a loose lower manifests itself as a Reverse C-bend in the mast. This is simply remedied by tightening the lower that faces the inside of the C until the mast is in column again. On a double spreader rig, this same problem will manifest itself as an S-bend as the lower will cause the bottom panel to lean to leeward, but not the intermediate panel. A reverse S-bend is caused by the intermediate being too loose and not the lower being too loose. This will be discussed further in the last chapter where we discuss tuning the rigging.

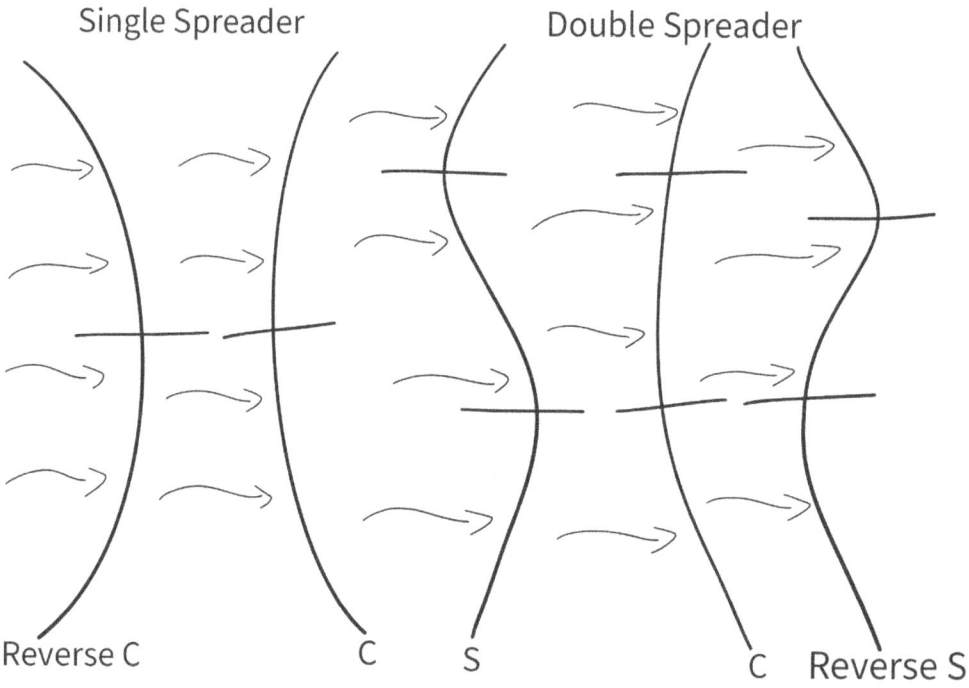

As you increase the number of spreaders, you also increase the number of possible rig tune issues you could possibly encounter, and as a result, increase the complexity of tuning your rigging.

There is yet one other issue with having more spreaders on a boat, and this is why most long-distance cruisers prefer a single spreader rig. If the cap shroud or intermediate shrouds break, the mast will snap at the first spreader when it comes crashing down. This is because the lowers are the only stay that reaches the mast at an effective angle without additional assistance, and is inherently the most stable part of the rig. When all else fails, lowers won't, and wherever they attach to the mast will remain as the rest of the mast comes crashing down. Very rarely, the whole mast will snap off at the deck, but that is usually caused by rather extreme circumstances—and at that point, the rig design doesn't matter because all of it is gone!

THE MAST ITSELF

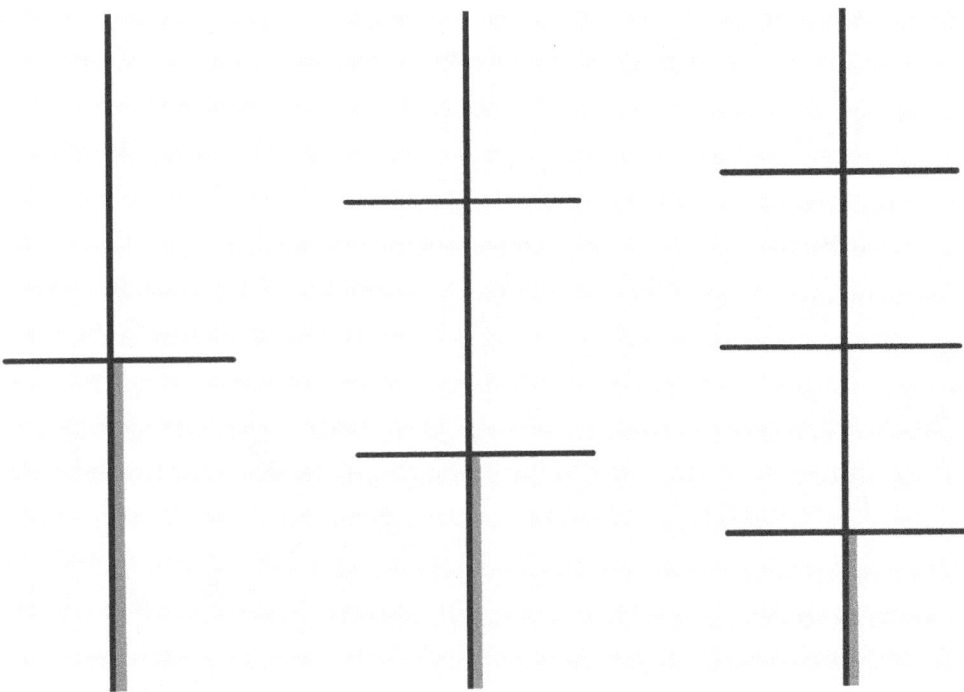

On a single spreader rig, the lowest panel is half of the original mast height. On a double spreader rig, the lowest panel is a third of the original mast, on a triple spreader rig, the lowest panel is a quarter of the original mast height. If your mast breaks at the lowest spreader, that means that the remaining mast you have left to get you home with a jury rig is only a half, a third, or a quarter of your original mast. Which would you want to have in that situation? For this reason, long-distance sailors who cruise to remote locations where help is nonexistent prefer a single spreader rig.

The single spreader rig is heavier and less finicky while adding spreaders, allowing for the rig to be made lighter and more adjustable. This is why racers abhor a single spreader rig and why cruisers with multi-spreader rigs suffer from their mast being out of tune.

While the spreaders help denote the amount of bendiness the mast will have, there is still another factor that plays a major part in this: How far up do the cap shrouds go?

On most traditional rigs, the cap shrouds are named so because they go to the top of the mast. This lets them act on the top of the mast and keep it from bending to leeward. As sail plans have become more radical—and on multihulls, which are incapable of heeling in a puff of wind—having a more flexible mast that would bend to spill the excess wind has become a desirable trait and the position of the cap shrouds has actually moved down on the mast. They are still called cap shrouds because they go up the highest, but they don't always go up to the top. Catamarans often have this setup as a safety feature, where if the wind builds too high, the top of the mast will spill the wind and help prevent the windward hull from coming out of the water or, worse, tipping over and capsizing. Some monohulls with square top mainsails have this feature as well; the top of the sail has so much area that if it were not to spill wind in a gust, the results could be catastrophic overall.

On these setups, the long unsupported length of the top part of the mast derives its rigidity from all the support offered below, making it relatively stiff and able to resist the pressures generated by the sails until they get too extreme and the masthead bends over to leeward just enough to spill the excess wind.

MAST SHAPE

When discussing wooden masts, we briefly touched on the shape of the mast section. A box-shaped mast is literally the shape of a box. It has four sides and is either square or rectangular. These are the easiest to make out of wood as they are merely four boards glued together. However, "easy to make" doesn't always mean "best possible," and as you can imagine, this is literally a box in the wind. The four main shapes of masts are square, round, oval, and aerofoil.

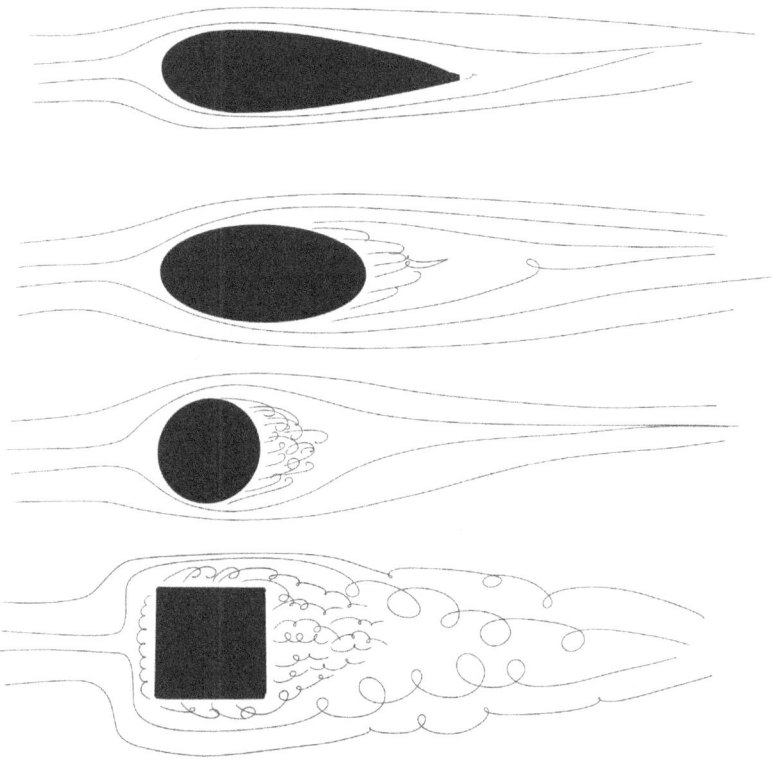

THE MAST ITSELF

Square

Square masts are characterized by having four flat sides. The corners can be sharp or rounded slightly, but the end result is still not very aerodynamic. As you can imagine, the leading face of a square mast is nothing more than a flat board in the wind. This shape is bad for a number of reasons, first being that the mast itself will generate a lot of drag, which detracts from the lift generated by the sail following it. The greater the drag force, the less power is available to propel a sailboat forward through the wind.

The second problem is it destroys the airflow over the mainsail. Imagine an airplane wing. Now imagine an airplane wing with a 2×4 stuck onto the front edge of the wing! When the wind hits a square mast, it is instantly pushed or pulled in a rapid and violent manner, which will create lots of spikes and drops in the local air pressure. The air at the beginning of the chord is some of the most critical in generating lift and this is the most disturbed section of the sail's airflow. This greatly reduces the total power that the sail could ever dream of making, which relates to a lower ability to power to windward.

A square mast with sharp corners is the worst as the sharp corners wreak havoc on the airflow. The corners can be rounded to attempt to smooth the airflow over the spar. This is a nice gesture, but the attempt is rather futile as the overall shape is still a brick in the wind.

Round

Rounding the corners to their absolute maximum will result in a round mast, which may seem like the answer since it is the polar opposite to a square mast, but such is not the case. A good way to compare the aerodynamic properties of shapes is to look at their coefficient of drag, or the number that reflects how much drag a shape will create; the bigger the number, the greater the drag and the more it will slow you down. While the coefficient of drag of a cube in the wind is about 1.05, the coefficient of drag of a sphere is merely 0.47, a vast improvement but it could still be better! Rounding the corners all the way down to create a circle-shaped mast profile greatly reduces the drag that the mast will generate, but it still doesn't help generate a smooth flow of air over the sail, which is the power plant of a sailboat.

The drag of the mast and amount of turbulence is lessened, but the large round spar will still create a large wake in the air where the sail will try to function. If we could keep the rounded shape but lessen the width, we might be able to improve the function of the sail while still keeping the drag down.

Oval

By flattening the sides of a round mast, you decrease the frontal surface area, which reduces how far the air molecules need to travel laterally to get around the mast. This decreases the drag somewhat, but, more importantly, it decreases the turbulent air that follows the mast onto the sail. Oval masts are significantly better at providing the sail with cleaner air than square or round masts, which means that the sail can generate more lift and produce more power to move the sailboat through the waves. Think of it as mashing the brake pedal down in your car while trying to drive down a road, then suddenly, you take your foot off the brake! Everything about the car is the same, but suddenly it can drive so much more quickly.

Aerofoil
If oval shapes were the best shape for aerodynamics, then airplane wings would be shaped like ovals. While the drag coefficient of a round mast at 0.47 sounds magnitudes better than the square shape at 1.05, both of these are very high when compared to a proper aerofoil with its coefficient of drag coming in at a whopping 0.04! This is all due to the way air is managed as it flows over the surface of the mast.

There are no sharp turns or harsh surfaces that will create turbulence as the air flows along the surface of an aerofoil, and, more importantly, the taper on the trailing edge is gradual. This keeps the air flowing along the surface of the foil, which improves flow by reducing turbulence and, as a result, reducing drag.

As you can imagine, a sail that receives clean and turbulence-free air will perform as well as it possibly could! The leading edge of the sail and the first part of its chord will have fast-flowing air that moves in uniformed patterns over the surface of the sail, which will generate as much lift as physically possible. The result is a very powerful system to support the sail and generates minimal drag.

This is great for thought experiments where the wind comes at the setup from the front, but the one direction a sailboat can never sail is directly into the wind! In reality, the wind will always be coming at an angle to the setup and the drag will never be as perfectly reduced as it could be.

FIXED OR MOVING MAST
Fixed Mast
The most common mast setup on a boat uses a fixed mast where the mast is firmly planted on the boat and stayed in place by all the standing rigging. There is little to no movement in the rig, as this part is supposed to be relatively static. The sails do their part to move about thanks to the constant adjustments of the running rigging.

While this setup is the simplest, if you are trying to maximize performance, then the turbulent air generated at various angles of wind will be most unsatisfactory. Imagine if your setup could be like the ideal drawing on p. 50. This can be accomplished—all you need to do is rotate the entire mast to face the wind!

THE MAST ITSELF

Rotating Mast
Rotating masts are stepped on a special type of mast step that allows the spar to pivot while under the immense compressive forces generated by both the standing rigging and the sails. These masts are able to rotate so that the leading edge of the mast perfectly meets the wind and creates the smoothest of airflow over the sail. The sail can perform its absolute best thanks to the abolishment of turbulent air as the mast itself is trimmed to the wind.

If you haven't already thought about it, now is a great time to pause and wonder: "What happens to the standing rigging when the mast rotates?" These rigs are very complicated and must be engineered in such a way that as the mast rotates, the length of the stay does not need to change while moving. By keeping the length the same, the tension of the stay is maintained and everything works out perfectly! Needless to say, unless your sailboat is a high-performance racing boat, it will not have this kind of complicated rig and will simply have a fixed mast instead.

Canting Mast
Taking the entire rig and rotating it wasn't complicated enough for performance-seekers, which is why they came up with yet another incredible advancement: canting masts. While the rotating mast turns to face the wind better, a canting mast leans to windward for many incredible performance benefits.

First of all, the sail is tipped to windward, which changes the way the wind interacts with it. As the sailboat heels to leeward, the sail is held upright and vertical, letting the sail perform its best in the most wind possible.

Having the mast lean to windward lets the sail actually act as a wing to help lift the entire boat up slightly. This lift helps to reduce the weight of the boat in the water, which for ultra-light high-performance boats can help them get up on plane and far exceed hull speed. (Hull speed is the "speed limit" for all displacement boats and is determined by the length of your waterline.) It used to be that the only way to sail faster was to sail with a longer boat, but by planing, any boat can exceed these limitations and sail faster on a smaller boat.

Canting rigs are incredibly specialized setups that so far only exist on boats designed to have them. You won't see an old boat cruising the Bahamas with a canting rig on it anytime soon.

WEIGHT ALOFT
When discussing the effects of the different mast materials, I made it seem like there were several tons of torque always trying to pull your boat over and capsize it, were it not for the ballast in the keel. While the force is up there, it is not exactly true that it is "always" trying to pull the boat over. The force of the weight aloft is dependent

on the angle that the weight sits in relation to the center of gravity of the boat. If the boat is vertical, the weight aloft is directly over the center of gravity and has no real effect on the matter. The weight aloft is merely a compressive force pushing straight down on the boat beneath it.

The value listed for the number of pound-feet that each section of mast length would induce on the boat is the force at 90° of heel, as this is when the force would be tangent to the arc of rotation. In other words, if you have a 1-pound weight at the end of a 10-foot pole that you are holding vertically, the weight of the entire setup would be just 1 pound. If you hold that same pole horizontal, suddenly the weight feels like it has increased dramatically to the tune of 10 pounds.

The weight of your spar at rest while the boat is anchored or in a marina is merely the weight of the spar. The pound-feet from all that weight aloft don't come into play until you start to heel over with the wind in your sails. The wind pressure will exert a heeling force on the boat, which is then compounded by the weight aloft leaning out over the side and increasing its effects of torque on the boat as it rotates around its axis of rotation. The more you heel, the more effect the weight aloft will have on pulling the boat even farther over, but at the same time, the more effect the ballast in the keel will have to bring the boat back to an upright position.

While this is not related to rigging at all, canting systems are not a new concept on a sailboat. For many years, racing sailboats have taken advantage of canting keels, which will kick the ballast to windward to increase the force and effect of the ballast to maintain the sailboat in a more upright position. This lets the yacht carry more sail and have those sails be more vertical in the wind, which greatly improves their ability to function and provide the propelling force to the sailboat below.

The type and placement of the mast on a sailboat is the most important decision, as this will have massive implications on the performance and sailing characteristics of the boat. What you intend to do with your boat is vastly different from what another sailor wants to do with their boat, and for this reason, it is so important that you choose the right mast, mast setup, and rig setup for what you plan on doing. If you simply buy a boat with the same rig that everyone else has, but plan on doing your own flavor of sailing, cruising, or racing, then you will probably have the wrong setup for your particular needs. By understanding all the variables and the relationship they have with all the other aspects of the mast, you will be able to make an informed choice about what you need to have, and how to use it to its fullest potential.

CHAPTER FIVE
How Running Rigging Ties In

STANDING RIGGING HOLDS THE SAILS IN THE BREEZE SO THAT THEY CAN PRODUCE the necessary power to propel the vessel forward, as well as transfer the power generated by the sails to the hull to pull the boat along with the sails. Running rigging is what gives the sails their power!

The curves of the sail are what is so important—they form the aerofoil that harnesses the wind. The leading edge of the aerofoil generates lift, which pulls the sail forward. There are a few theories about how this works, which means we really don't know how it works for sure, but since it does work we can all go sailing. In chapter 6, we will go over these competing theories and let you decide which you like the most. But before we get there, let's look at how running rigging controls where the curves exist on the sail.

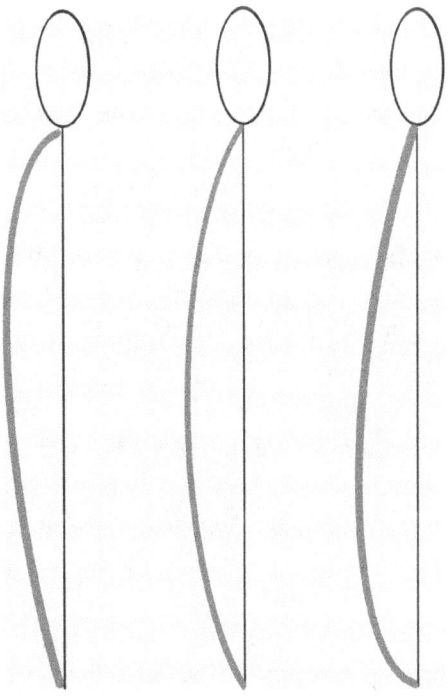

First, we need to go over the terminology of a sail as an aerofoil. The total length of the sail luff to leech is called the chord, while the belly that the sail develops is called the draft. As the draft of the sail deepens, so does the power that the sail can generate. As the draft changes, so does the angle of attack and this affects how high upwind the boat can point.

Having the draft in the middle of the sail will generate the most power, but if the wind builds, moving the draft forward will help reduce some of the power in the sail and let you continue sailing comfortably with the same size of sail flying in the air. If the draft moves towards the leech of the sail, this will cause the boat to heel more as the drag from the sail increases dramatically. This increased drag actually slows the boat down while pulling it over harder.

HALYARD TENSION FOR UPWIND SAILING

The halyard does more than just raise up a sail; it also provides necessary tension to the luff of the sail, which causes the draft to move forward on the chord. Standard procedure for raising a sail is to pull it up until the head of the sail is as far up as it can go and the luff becomes tight, then crank on the winch until vertical wrinkles develop parallel to the luff of the sail. Depending on the wind conditions present, you might want to change the halyard tension to better suit your power needs from the sail.

In light winds, you want to move the draft aft to the middle of the chord, which is accomplished by reducing halyard tension. In heavier winds, you want to move the draft forward to depower the sail slightly, which is accomplished by increasing halyard tension.

As wind speed changes, so does the power it imparts upon the sail, as does the amount of stretch the sailcloth experiences. Setting the halyard nice and tight where the draft is right behind the mast at 15 knots of wind does not mean that the draft will still be forward of the middle when it is blowing 30 knots. As the wind increases, the pressure on the sail does too and the draft will continue to move farther aft than it was in lighter winds.

When it becomes physically impossible or impractical to crank the halyard winch tighter, the Cunningham comes into play, providing more mechanical advantage to tighten the luff and move the draft forward on the sail. As mentioned in chapter 2, the Cunningham attaches about a foot above the tack and uses a block and tackle setup to tighten the luff even further. Some sailors prefer to use the Cunningham to adjust luff tension because of the increased control and immediate response provided by the sail. Another reason it is well loved is it negates the need for a halyard winch on a very small yacht. The mainsail can be raised by hand and cleated off as tight as easily possible, then the Cunningham simply takes any slack out of the sail by pulling down on the luff while setting the draft in the perfect location.

Adjusting the luff tension is also useful with the headsails, as these are often overlooked since they are not as noticeable, lacking a mast and a vertical luff. Headsail draft position is best seen by the aid of draft stripes, which are contrasting color stripes to the sailcloth. These stripes are useful to visualize the curve of the sail and visually represent the position in the sail where the draft is at its greatest. Cruisers probably haven't adjusted the halyard tension of their furling headsail in years and don't really worry about where the draft is located. The problem with ignoring the headsail luff tension is over time the halyard will stretch and the luff tension will diminish, causing

HOW RUNNING RIGGING TIES IN

the sail to be a bit more baggy. The result is a less comfortable sail as you heel more, which will cause the cruiser to furl away some of their headsail to reduce the sails power. The smaller sail will be less effective than the full sail as the partially furled headsail creates a lot of turbulent air, just like a round mast does. All of this is easily avoided by simply checking how tight the halyard is and giving the winch a crank every so often to check for any stretch that may have happened over time.

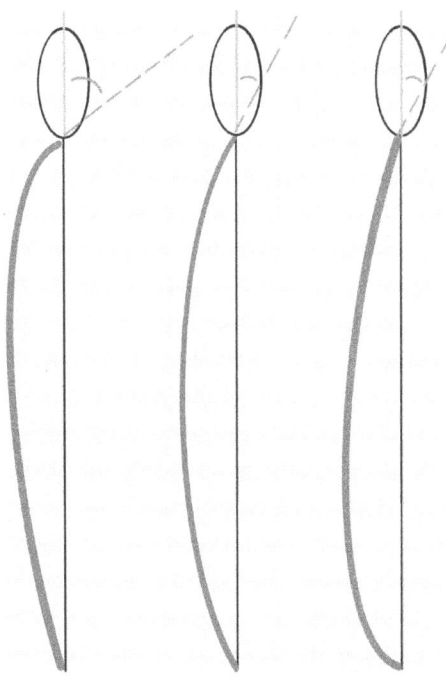

The position of the draft has another major implication for sailboats, it dictates how close you can sail to the wind. Ideally, you want the wind to flow onto the sail undisturbed and then be guided along the curve of the sailcloth to generate lift. If the change in angle is too abrupt, the wind will separate from the surface and result in turbulent air that creates drag instead of lift.

The angle that the sail, acting as an aerofoil, meets the wind is called the angle of attack. Too sharp of an angle and the sail will stall, losing all power and acting as a drag device in the wind, pulling the sailboat over to leeward. To control the angle of attack, you can trim the sail in or out, but you can also change the angle of the leading edge of the sail by moving the draft fore or aft. As the draft moves forward, the leading edge of the sail becomes deeper and this means that the sailboat will not be able to point as high into the wind. Moving the draft back flattens out the leading edge, which allows the sailboat to point higher into the wind.

As the sailboat moves faster, the apparent wind speed will increase and so will the wind pressure on the sails. This will undoubtedly cause the sailcloth to stretch slightly, which will cause the draft to move even farther aft. This is why halyard tension is far from a "set it and forget it" situation as you need to pay attention to the draft and the relationship with luff tension as conditions change.

HALYARD TENSION FOR DOWNWIND SAILING

For sailing downwind, the sail relies more on drag to pull the sail along rather than lift. If increasing halyard tension increases luff tension, which decreases drag, shouldn't the opposite be true for downwind sailing?

By easing the halyard tension, the luff tension disappears and the sail will become baggier. This will fill with more wind and pull you downwind with great speeds! Vertical creases and wrinkles are the sign to look for when setting a tight luff, but for a loose luff you want to see horizontal creases and wrinkles form between the hanks or slides. These horizontal clues will tell you that the sail is up but loose, and when the wind blows on it, it will turn into a giant bag in the wind. It is important to note that in light winds, you will see the horizontal creases form between the hanks or luff slides, but when the wind picks up you will see the sailcloth creasing in a V shape emanating from each hank or luff slide. If you see this happen, you are stressing the small cringles on the luff of the sail and pushing them towards their breaking point. It is advisable to add halyard tension until these V-shaped creases disappear and the force of the luff is split between the halyard above and the tack below, with the hanks or slides merely guiding the luff of the sail.

If you are sailing downwind and want to have a nice baggy headsail pulling you along but you don't want to tear out the cringles from your luff, you can do yet another trick. Tighten the halyard until the V-shaped creases disappear but then ease the backstay to relax the headstay. While the luff of the sail is tight and the hanks are merely guiding the sail to conform to the shape of the headstay, the headstay itself will sag and bow out to leeward. As a result, you will have a headsail with proper luff tension following along on a curved path dictated by the slack headstay. This bowing is called headstay sag and is very useful for downwind sailing as it causes the headsail to fill deeper and generate greater power.

OUTHAUL TENSION

While halyard tension is used to adjust luff tension and position the draft, outhaul tension dictates the size and depth of the draft. Easing the outhaul will allow the sail to have a deeper draft, which will generate more power, while tightening the outhaul will flatten the sail and depower the sail at the same time.

In lighter winds, it is advisable to ease the outhaul as far as you can so that the sail will have as deep a draft as possible. This will generate as much power as the sail possibly can in the light winds, which hopefully is enough to make the sailboat move through the water. As you know, a deeper draft results in a greater angle of attack, which means that the sail won't be able to point as high into the wind.

As wind speed builds, so will the power generated by the sail. When it reaches the point where no more power is wanted from the sail, you can either change to a smaller sail, which will produce less power because of its lesser surface area, or you can depower the sail that is presently set. Tightening the outhaul will pull the clew of the sail aft and flatten the sail, resulting in a shallower draft and less power generated

HOW RUNNING RIGGING TIES IN

from the sail. The shallower draft also decreases the angle of attack, which will allow the sail to point higher into the wind, but while developing less power.

Using the outhaul in combination with the halyard or Cunningham is a powerful combination as you can control the size and position of the draft to power or depower the sail as you need. You can also control the angle of attack for upwind sailing and turn your sail into an efficient drag device for speedy downwind sailing.

TOPPING LIFT

The most forgotten piece of running rigging is the topping lift, which is easily identified on any boat by the surface algae growing on it. The purpose of the topping lift is to support the boom when the mainsail is lowered. If there were no topping lift present, the boom would only lift when the leech of the sail was tight enough to lift the boom along with the clew. In operations such as raising the main or reefing, the boom would fall into the cockpit and smack everyone in there.

The topping lift is normally set to hold the boom just clearing the top of the bimini and dodger. This has since led to a "Set it and forget it" attitude toward the topping lift.

The problem with a slack topping lift becomes even more problematic when the mainsail is reefed, as the reefed clew is raised, which in turn lifts the boom higher and makes the topping lift even more slack as it bows out in the violent wind. For sailors that like to tie their ensign to their backstay, the whipping and wrapping of the topping lift around the backstay can quickly escalate to a horrible situation in which the flag seizes and holds the topping lift captive.

Imagine tacking through the wind in a violent storm from one beam reach to the other, only to find that your topping lift has seized itself to the backstay and now you are forced to close haul directly into the wind and waves for the remainder of the storm until someone climbs onto the end of the boom to free the topping lift.

The easiest way to prevent all of this catastrophe is to simply pull in on the topping lift anytime the mainsail is raised or changed. This will keep it tight and mitigate any risk of the topping lift fouling the backstay.

On a much more pleasant note, the topping lift can also be used for downwind sailing, especially in tight waterways such as rivers or bays, where you are forced to sail on a dead run and might need to change course at a moment's notice to avoid a jet skier or clueless powerboat. The trick is to over-tighten your topping lift, causing it to lift the boom a few feet up. This will make the leech of the sail to twist uncontrollably, which in normal situations is a massive problem but in this situation is exactly what you want it to do. Once the boom has been raised, you will sheet the main in tight so that the boom rests over the centerline of the boat but the top of the sail with its twist lays perpendicular to the wind.

The bottom third of your sail is worthless in this setup, but the top two-thirds of your sail will lay out into the wind and drive the boat forward on a run. If you need to jibe, the boom is incapable of swinging or slamming across the boat, as it is tied in tight midship, but the top of the sail will gently flop over to the other tack as effortlessly as a self-tacking headsail.

This trick is really handy when sailing the Intracoastal Waterway (ICW) of the United States, as the waterway is only as wide as the channel, and all the buildings that line the waterway as you pass through cities will cause some strange windshifts. Trying to sort out the changing wind as well as staying in the channel can prove too much for sailors and cause them to either crash jibe or run aground. Trimming the mainsail like this takes all the guesswork out of the situation and lets you focus on staying in the channel while the wind is at your back.

SHEET POSITIONING

The first way that comes to mind to adjust the sheets would be at the winch, either letting the sheet out or cranking in on the sheets. While this is true, it is only part of the way the sheets can be adjusted.

Where the sheet pulls from is just as important as how much the sheet pulls. The clew of the sail will always follow the direction and angle that the sheet is run, and this position is controlled by the position of the sheet block.

The "neutral" position for the sheet block is when the line created by the sheet carried forward onto the sail as an imaginary line meets the luff at the 40% height point, also called the luff perpendicular. At this position, the sheet is not exactly bisecting the headsail, but instead pulling a smidge more on the foot than on the leech. This should always be considered the starting point for the sheet block, as this is the neutral position.

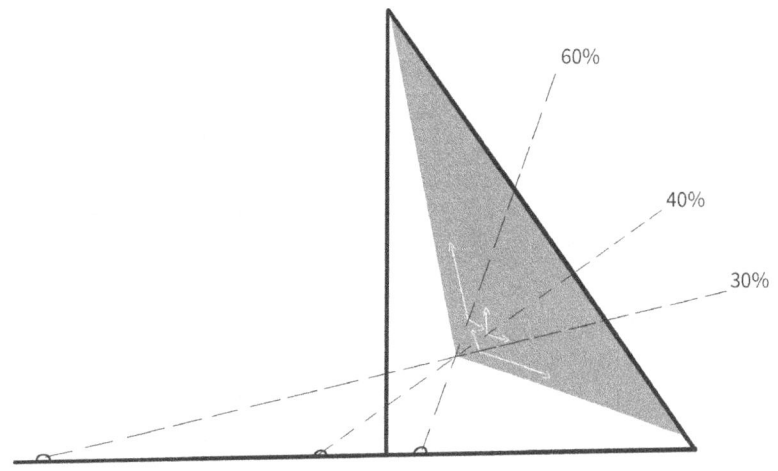

HOW RUNNING RIGGING TIES IN

Moving the block aft will change the sheeting distance and move the force onto the foot of the sail. This will flatten the sail out and improve the control in heavy winds where a flatter sail will not generate as much power. Less power from so much wind will still be plenty of power! It is like trying to fill a glass with water: If you are filling it from a galley faucet, you want to catch all the water there is and not waste any of it; but if you are filling it from a fire hydrant, you will do much better if you just skim a little bit of water off the surface.

In light winds, just like with the galley faucet, you want to harness all of it and not let any slip by you. Moving the block forward will transition the force from the sheet to the leech of the sail. This will relax the foot, which will cause the sail to generate a deeper draft and more power.

Moving the block fore and aft seems like a pretty common adjustment on a sailboat, as wind conditions will vary greatly in a single day. For this reason, the sheet block is normally mounted on a track that runs fore-aft on the deck. The setup is called the jib track and the block is called the jib car.

While most boats have the jib track and car setup, some boats do not for a variety of reasons, most notably when they have a self-tacking jib and the jib track and car run athwartship in a setup more akin to a mainsheet traveler. The method employed to adjust the fore-aft sheet angle is to use a clew board that simply has a series of holes arranged at the clew instead of one single hole. Moving the sheet attachment up the clew board is akin to moving the sheet block forward, as the pull will be directed more into the leech of the sail and is best for lighter winds. Moving the sheet attachment down the clew board is akin to moving the sheet block aft, as the pull will be directed more into the foot of the sail and is best for heavier winds.

Having a long string of holes on both sides of the deck for the jib track bolts gives you the ease of trimming your sail to the utmost perfection, but it also invites a lot of leaks into the deck of the boat. The big advantage of a clew board over a jib track is the reduction in the number of holes in the deck. The clew board method only has around five holes, so you don't really get that much selection, but the jib track can be replaced by a single padeye on each side of the deck, greatly reducing the number of holes but also increasing the load at that very site, as you have fewer fasteners and a smaller area where the force is being distributed to. As with everything on a boat, it is always a compromise.

All of this covers the sheeting position of a sail in a fore-aft motion, assuming that the athwartship position remains fixed, but this is not the case. First of all, the positioning of the jib tracks was carefully selected by the naval architect based on the sheeting angles desired for the boat. The closer the sheets can pull to the midline of the boat, the closer the sails can be sheeted into the boat and the higher the boat can point into the wind.

THE RIGGING HANDBOOK

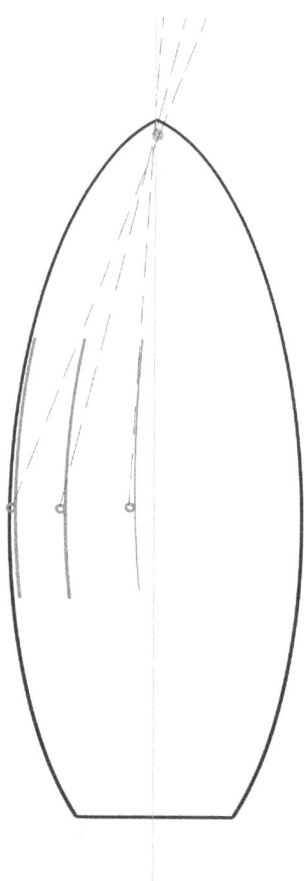

The clew will always be pulled to the sheet block. This means that if the sheet block is on the toerail, the clew will be pulled to the toerail. This limits how far upwind the boat can point, as the sail needs to maintain a proper angle of attack to the wind, which cannot exceed the angle the sail has to the wind. If the clew of the sail could be moved inboard, the angle of attack will move with it and the sailboat can now point higher into the wind.

On some boats, the sheet block is infinitely adjustable in all directions and dimensions, since it is merely a frictionless ring suspended by the tensions of various control lines. This allows the ring to move fore-aft, as well as inboard and outboard. For us mere mortals with boring jib tracks, we might feel unable to achieve these grand arrangements, but such is not the case. Barberhauls and inhauls are the answer for us regular sailors who wish to improve the performance of our sails on extreme sheeting angles.

Barberhauls
When sailing downwind, the sheet is eased and, as a result, so are both the foot and the leech, causing them to twist uncontrollably, spilling the wind they are trying to harness, costing you performance. Worse yet, the leech and foot can flutter, which will work harden the yarns of the sailcloth as well as work out the resin that holds the sailcloth yarns together, resulting in the premature death of your expensive sail.

The solution is to find a way to sheet in the sail while still having it positioned to leeward. This is where the barberhaul comes into play. By definition, a barberhaul is any block arrangement that moves the sheeting angle outboard. This can be accomplished by using an infinitely adjustable frictionless ring that is suspended by various control lines or by simply mounting a snatch block on the toerail between the jib track car and the clew. The sail and sheet have no concept of the path the sheet takes once it passes the last sheet block, so repositioning the block to the toerail will effectively open the sheeting angle outboard. Having the block mounted somewhere near the clew of the headsail will allow you to sheet the sail in tight to tighten the leech and foot of the sail while still at an angle that is more conducive to an off-the-wind course.

HOW RUNNING RIGGING TIES IN

Inhaul

When sailing upwind, the normal sheet block position might not be close enough inboard for specific purposes. The upwind sailing angle of the boat is dictated by how far inboard you can sheet your sails. This is within reason, as you can't sail straight into the wind by sheeting your sails to the centerline of the hull. To pinch a bit farther upwind, you will need to close the sheeting angle and move the sheet block closer to the midline of the hull, sheeting the sail closer to the midline of the boat. While a barberhaul relocates the sheet block outboard, an inhaul does the opposite and brings the sheet block inboard. On racing sailboats, this setup is commonly used with purpose-cut blade jibs that are upwind beasts.

The same holds true with someone sailing on an older cruising boat where the jib tracks are located on the toerail to "free up the deck." As a result, these boats will find it impossible to sail close to the wind simply because their sails cannot be sheeted in all the way. Adding an inhaul to this setup will be of a huge advantage, as you are now able to reduce the sheeting angle and sheet the sails in over the deck and allow the sails to be trimmed for close-hauled sailing.

REEF LINES

If there is one type of line on a boat that no one thinks about making sure it is set up properly before going sailing, it is the reef line. The time to address issues with these lines is well before you need them, because once you start thinking about needing them, the last thing you want to do is spend time untangling or running them.

Reef lines are a rather broad category that carry out the task of making your sails smaller. With roller furling, these lines are becoming less and less known; roller furling allows for what is called "roller reefing," where you simply furl up the unwanted sail area and gradually make the sail smaller.

While roller-furling headsails have pretty much made sightings of headsail reef lines on par with sightings of the Loch Ness Monster, mainsail reefing is still commonplace by those who want better performance out of their sails.

The purpose of reef lines is to replace the standard tack and clew with a new tack and clew that is closer to the head of the sail. This effectively makes the whole sail smaller as the foot of the sail is removed from the wind. In reality, the sail actually comes down as the luff length is decreased, moving the center of effort of the sail lower so it has a shorter lever arm to heel the boat over in high winds.

On a mainsail, there are reef points sewn into the sail. The luff will have the tack reef point, and the leech will have the clew reef point. These reef points are built into the sail by the sailmaker and will correspond to the number of reef points that your sail has. Most daysailers might not have a reef point at all; if they do, they probably

only have one. Most coastal cruisers will have two reef points, while ocean-cruising boats should have three reef points.

When reefing, the tack reef point becomes the new tack and the clew reef point becomes the new clew. This makes the sail smaller, and instead of being connected to the rig by the tack shackle and boom car, it is connected by ropes or hooks and straps.

The tack point is usually accomplished with a simple hook that provides a very rigid point of attachment for the new tack and holds everything as well as the normal tack shackle did. The tension on the clew reef line has the same effect as the tension on the outhaul, pulling the sail flatter to reduce the draft as wind builds. The sail is smaller and has less potential to grab the power from the wind, but there is so much more power to grab that a smaller sail is plenty for the boat to sail comfortably.

The clew reef line, which acts as the new outhaul, is a standard line on every mainsail that doesn't have roller furling, either in-mast or in-boom, and has the ability to reef.

The tack reefing is less of a standardized system. Tack reef points can be secured by a metal reefing hook, or by a variety of line arrangements via a single reef-line system or a double reef-line system.

Single Line Reefing

Single reef-line systems seem the most intuitive, as they make the entire process of reefing seem much simpler to comprehend and carry out. Reefing is reduced to two lines normally controlled from the cockpit, which means you don't have to go on deck in snotty conditions as the wind starts to rip the tops of waves off and sting them against your skin! Reefing is simply easing the halyard while winching in on the "reef line." Once the reef line has been taken in all the way and the foot of the sail is as tight as possible, the halyard can be re-tightened to raise the sail back up to its new, shorter size.

This honestly sounds like a great method, and you might be wondering why reefing is done in any other form since this setup must be absolute perfection. The reason is because this one sounds simple but is actually rather horrible. The first problem is the absolute length of line needed for the sail to reef. If your reef point is 1 foot above

HOW RUNNING RIGGING TIES IN

the foot of the sail, then you now have 4 feet of line to winch in to accomplish the reef. This is because the line has to run from the cockpit to the mast, up the mast and to the gooseneck, up the luff to the cringle, then back down to the boom on the other side, along the length of the boom to the reef clew block, up to the cringle and back down to the boom. Whatever the height of the reef point is, you now need to multiply it by 4 and that is how much line you need to bring in. Most reef points are at least 3 feet above the foot of the sail, which means you need to pull in a total of 12 feet of line at a minimum for the first reef. As you go successively deeper into the reefing, the reef points get taller and taller, and that means the lines get longer and longer.

The standard placement for reef points is every 12% of the luff length. This means that the first reef is placed 12% of the way up the luff, leaving you with about 88% of the sail height remaining. The second reef is placed 24% of the way up the luff, leaving you with 76% of the mainsail's original height. The third reef is placed 36% of the way up the luff, leaving you with 64% of the mainsail's original height. If your mainsail is 40 feet tall, the reef points would be located at 4.8 feet (1.5m), 9.6 feet (3m), and 14.4 feet (4.4m) up the luff. The reef-line length that you would need to pull in would be 19 feet for the first reef, 38 feet for the second reef, and 57 feet for the third reef.

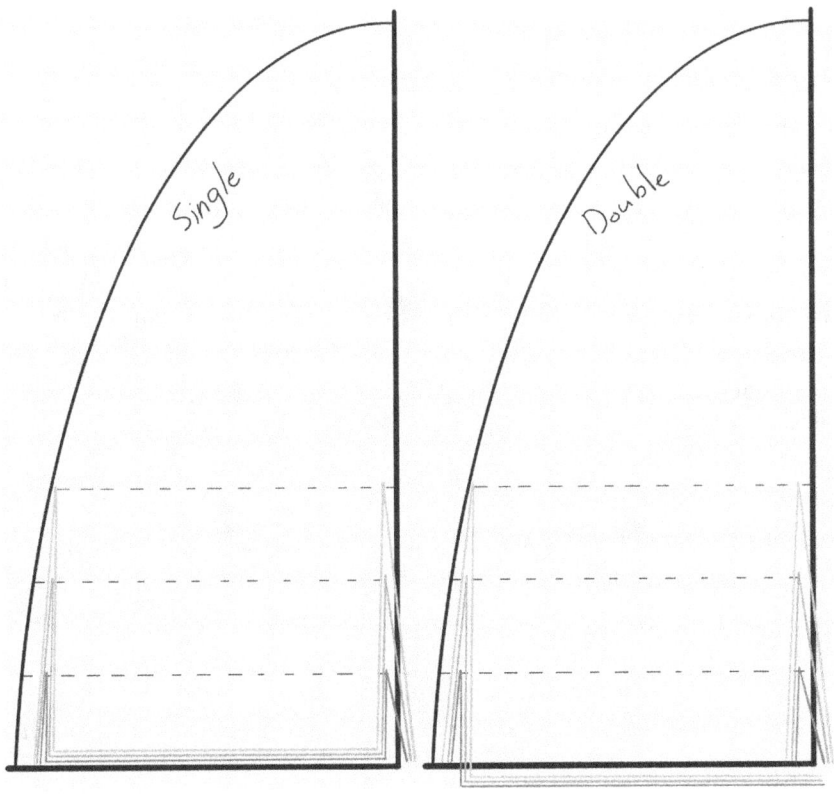

In case you think that pulling in almost 60 feet of rope is a fun activity, it is important to remember that all this length of line will only flow if it is not tangled, and as the wind begins to build, lines will twirl and twist around everything, even themselves. The result is a horribly tangled mess that may require you to now go up onto the deck and climb onto the boom to untangle the twisted lines in a violent windstorm.

The proper way to reef with a single line reefing system is to lower the halyard while simultaneously pulling in on all the reef lines to keep them taut and prevent twists. If you only have a single reef point, you now have two hands working two different lines. If you have three reef points, you now have two hands working four different lines!

Aside from the logistical problems of working a single line reefing system, there is one other major flaw in the system: tension. The tension supplied to the tack and to the clew is the tension supplied by their respective lines. These parts of the sail pull in different directions and under different loads, but when they share a line, they also have to share tension as well. A happy medium between the two parts of the sail does not result in a happy sailor. The goal is to have enough tack tension to hold the tack close to the mast to take all the force of the sail being pulled backwards, and if the tack doesn't have enough tension, the first sail slide cringle above the reef point will tear out of the luff. The clew needs a lot of tension to flatten the sail and pull itself down to the boom; otherwise, the sail will have too much draft and become overpowered.

In theory, if you tighten the reef line to the point where the clew is appropriately tensioned, then the tack will also have enough tension. This is true, but the practical problem is each turn the line makes adds resistance, which then makes it harder to tighten. Think about when you are docking. Imagine you are standing on the deck of your boat and grab a line from shore; then the captain thrusts to spring off of that line. What will happen to you? You will get pulled off the deck! If you pass that line under the horn of a cleat, suddenly you can hold the line and the captain can torque against the line to safely dock the boat. If you give the line a full wrap around the cleat, you barely need to hold the line to keep it from slipping. The same holds true with single line reefing: Every turn that the line makes adds resistance to the system, and that means it feels like you are really cranking the line in hard when in reality the tack and clew are billowing in the breeze and won't tighten down that last bit!

Sailmakers have tried to remove resistance from the process by sewing blocks to the sail instead of just using a cringle, but this still doesn't fix the issue of line jams and tangles that can occur if a loose line was allowed to twirl and twist. Single line reefing seems like a great idea on paper, but if it were that great in reality, it would be standard equipment on all sailboats.

Double Line Reefing

Single line reefing had the right idea, but it failed in the execution due to having too long of a line and too much resistance in the line. Breaking up the two tasks at hand seems like a worthy compromise to fix the problem, and it does.

Reefing a double reef-line system is straightforward: loosen the halyard to lower the sail to the reef point while tightening the tack line. Next, tighten the halyard to raise the mainsail back up and to tighten the luff, then, tighten the clew reef line until the clew is nice and tight. Done! I have seen this procedure performed on a Pogo 40 by one experienced sailor with the help of one inexperienced sailor; they managed to reef the whole mainsail in about 20 seconds. Everything can be done from the cockpit and, more importantly, each line has less length that needs to be pulled with fewer turns generating less resistance. The process is fast and efficient, with little chance for lines to get irreversibly tangled.

As with all reefing systems that use a line for the tack, it is very important that the blocks for the tack be located on the mast, forward and below the tack point—that way, the force on the tack is both forward and down. If the tack is pulled only down, it will be pulled back and tear out the first luff cringle. If it is only pulled forward, the halyard will raise the tack up and position the sail outside of its desired placement on the rig.

For both of these systems, reefing is performed in the cockpit because the lines are all led aft. If you have a boat where the lines are kept at the mast, then you will need to go forward to the mast to carry out the reefing procedure. In these setups, the halyard and clew reef lines are located at the mast where one person can carry out the procedure. Setups where a tack horn is employed often also have the lines stay at the mast. Since you need to go forward to hook on the tack, you might as well do the rest while you are up there. The main disadvantage of this setup is that you need to leave the cockpit to go forward during a storm, but the main advantage is that everything is simpler. The lines simply run to their respective stations and terminate at the mast; there is no more routing or turning that they need to do, which greatly reduces the resistance on the line. Also, if a problem were to occur, it would occur at the mast. While you're already up there, you can see issues begin to develop and correct them before something breaks. If you are in the cockpit under the dodger looking down while cranking on the winch, you might not realize something was stuck until you hear a loud rrrRRRRIIIIIIIPPPPPPPPP!!!!!

Typically on coastal cruisers, double reef-line systems prevail (some have single reef-line systems) while on older bluewater cruisers, the lines are kept at the mast.

Going forward may seem frightening at first when you spend all your time in the cockpit, but everything depends on what you are used to. If you know you will never need to leave the cockpit until the anchor is down again, then you might not have the deck setup to make it safe to go forward while underway. If you know you will be running around on deck raising sails or reefing, then you will have both the experience of moving around on deck as well as the necessary safety equipment setup to make this a less-frightful task.

There are some boats where the lines are led aft but still use a tack horn and this setup is the worst. The halyard can only be controlled in the cockpit but you have to go to the mast to attach the reefing hook. If you are alone, you are facing an impossible task where you need to be in two places at the same time. If you have this setup, consider converting the tack to a line, soft shackle, or moving all the lines to the mast and abandoning the cockpit arrangement. This setup works well with two people, but if you are crew of two, and one of you gets incapacitated, you are now singlehanding and the boat should be set up to allow such a maneuver.

We have discussed tack horns as well as tack line setups, but there is still yet another alternative that is not popular at all, and that is the soft shackle. On my own personal boat, I had the tack horn system but got tired of doing luff repairs where the horn would hook the luff of the sail and tear a nice big gash as the sail

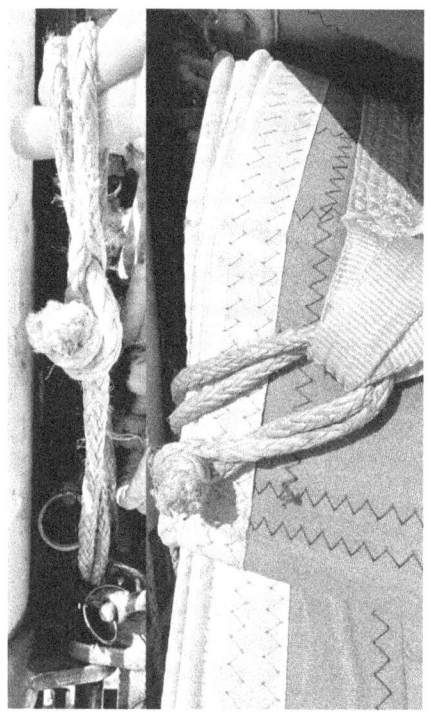

was raised, so I replaced the tack horn with a tack soft shackle. This method cannot cause any harm or damage to the sail while it is being raised, but it does not provide any forward pulling tension as it only pulls straight down, so it is imperative that the halyard be tightened before the clew is tightened. By tightening the halyard first, the luff is pulled tight and the soft shackle works vertically. At this point there is too much tension established for the tack to now be pulled aft so it remains vertical and the luff cringles do not rip out. If you tighten the clew before the halyard, the tack will slide aft and the luff cringle will tear out of the mainsail. This method is not foolproof, but it is the method I use on my own boat because, as long as you do it right, it will not harm the sail and is the simplest setup to cause the least amount of tangles.

HOW RUNNING RIGGING TIES IN

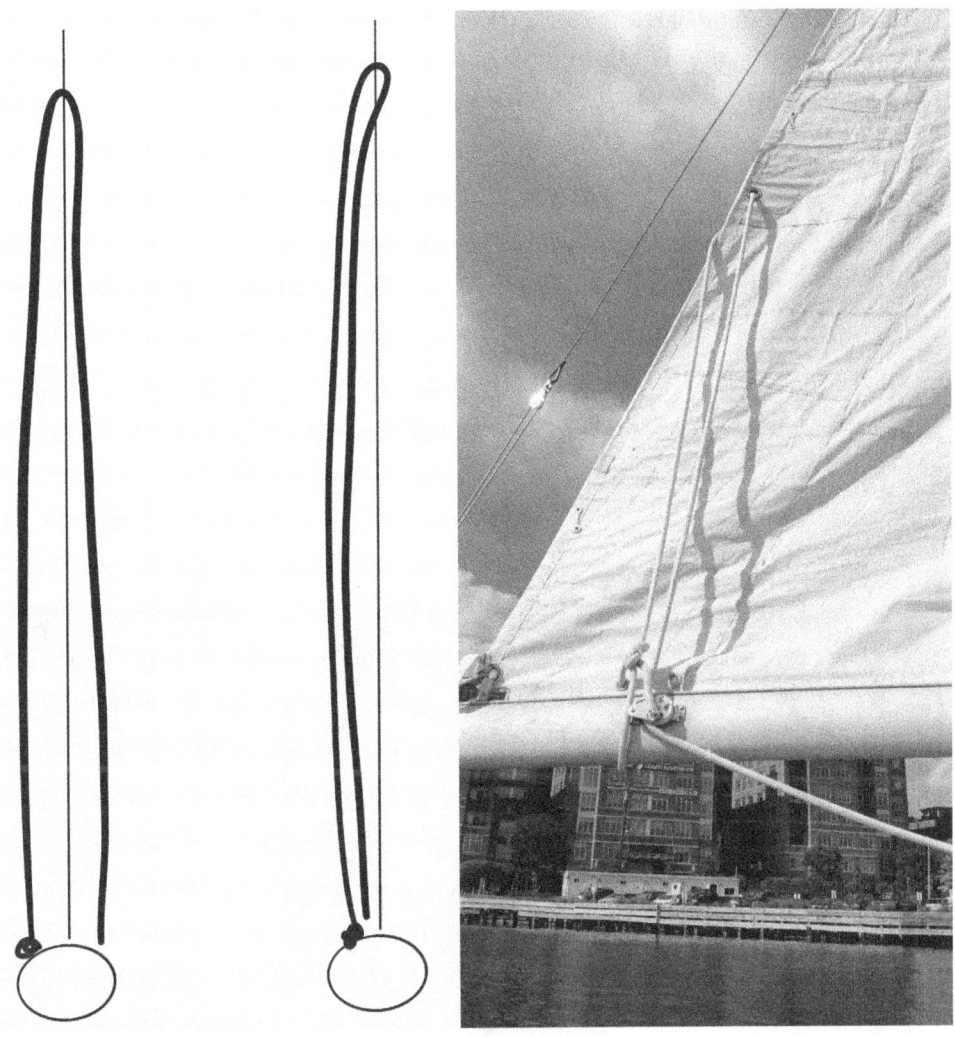

The standard way to run the reef lines at the clew is for the end of the reef line to be tied to the boom, then up to the clew cringle, and back down the other side. This creates a sling that holds the reef clew in place and pulls all the sail cloth down inside the sling. The problem is as your reefs get deeper, the amount of sail cloth that needs to cram in here increases and so does the likelihood that the cloth will get pinched and torn by the reef line. An alternative method is to have the reef lines be on the same side of the sail so that the sail is booted out to the side, allowing you to reef deeply without any concern about how much sailcloth is involved. This method starts off the same where the reef line is tied off on the boom, goes up to the cringle and through it, then it comes over the leech to come back down. This means that the ascending and descending reef lines are visible on one side of the sail but not on the other, whereas the traditional way has the ascending reef line on one side of the sail and the descending reef line on the other side of the sail.

FURLING LINES

While mainsail reefing using reef points still exists on most boats, headsail reefing has all but been replaced by the growth in popularity of roller furling. This means the furling line can be viewed as a modern and simplified reefing line, a true "one-line reefing system" that actually works!

Furling lines are small lines that run from the cockpit forward to the furling drum. They work by simply wrapping around the furling drum to let the sail out, and pull the furling drum along to wrap the sail around a tube that spins around the headstay.

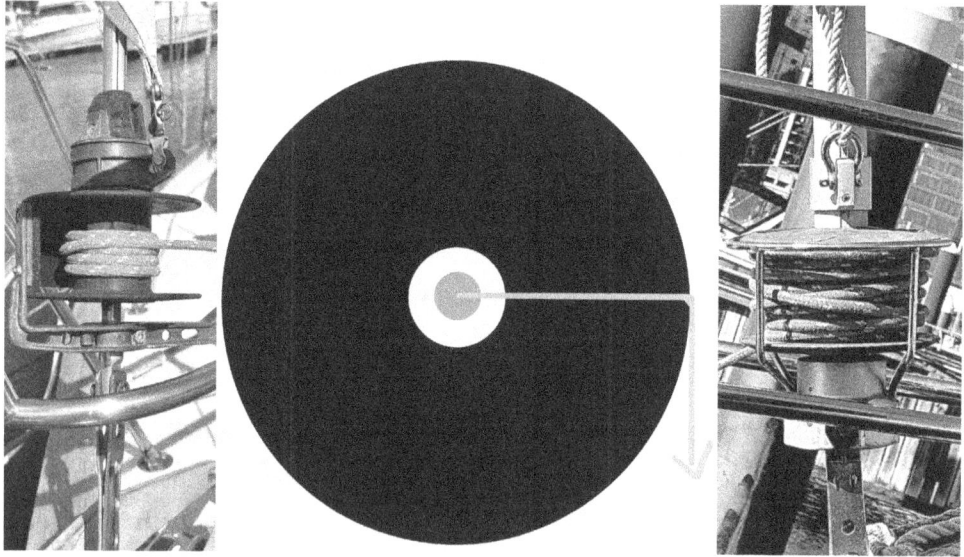

The most basic principle of the matter is that the furling system operates via a leverage system where the size of the drum dictates the lever arm that acts on the sail. The larger the drum, the larger its radius and the longer the lever arm.

As the furling line is pulled, the sail is rolled up, which reduces the sail area. Unfurling the sail is easy to do as the wind will catch the sail and if the furling line has no tension on it, the drum will spin quickly to release the sail. Furlers are a great convenience item on sailboats, which is why they have become so popular and appear on almost every kind.

HOW RUNNING RIGGING TIES IN

The greatest convenience is that you will never need to raise or bag a sail again, since the sail lives rolled around the stay and when you furl it away, it stows itself. This makes coming into port much more graceful; the sail is simply rolled out of the way, lifting its sheets up off the deck with it.

While roller furling is wonderful when it works, there are a few bad habits that are also popular and lead to a bunch of problems with furling. The first bad habit is releasing the furling line and letting the sail explode into service. We have all seen sailboats do this, and I must admit, it is very fun! You release the furling line and crank in on the genoa sheet. As the sail starts to come out, you crank in faster on the sheet until the sail catches the wind and bursts out as the boat gets pulled to leeward and you instantly have a wake behind you. This is about the equivalent to slamming on the gas pedal the moment the light turns green at an intersection; you shoot out and feel the rush of acceleration, but for what purpose? The additional wear and tear on your car is unjustified as you come up on the next light and it's red. The same holds true with your sailboat.

As the wind snaps the headsail into shape, this shock load is transmitted through your rig. The sudden stop when the sail reaches its end snaps the furling drum around, which can damage the internal bearings and races. Worst of all, the furling line rushed into the drum as the sail shot out, and this can lead to wonky and bulky wrapping. The best-case scenario is that the furling line will become jammed against the drum's cage, but the worst-case scenario is that the furling line develops an override, a situation where a line passes under itself and gets pinched, preventing it from unwinding later. When you go to furl the sail in as the wind builds, you suddenly can't because the furling line is stuck!

The correct way to avoid making this first mistake is to put the furling line on a winch with a few turns. As you go cranking in the sheets to pull the sail out and as the sail begins to catch the wind, simply ease the furling line on the winch in the cockpit so that the furling line slowly and tightly wraps around the furling drum in a controlled manner. This means the furling line is neatly wrapped on the drum and as a result will neatly come off when the sail is furled back in.

The second bad habit is not furling the headsail in all the way. Some people like to leave a little triangle of headsail showing from their furled headsail, but this puts all the stress on the furling line. As wind builds, the little triangle of headsail will catch the breeze and stress the furling line. If the furling line were to come loose—or worse, break—the headsail will quickly unfurl and deploy the full headsail into the wind at the worst possible time. If no one is around to fix this problem, the sail will flog itself to death in the wind and turn itself into shredded tassels by the time you return to your boat. The best way to remedy this is to simply furl in the headsail until the sheets do a minimum of two wraps—three is better—around the furled sail. If the furling line were to stop holding the sail, hopefully the wrapped sheets will help keep it furled up and prevent catastrophe until it is repaired. If you can reach the clew, tying a sail tie around the furled sail will also safeguard against the sail accidentally opening up while being stowed.

The third bad habit is forgetting to adjust halyard tension based on the conditions at hand. The halyard will creep over time when a sail is raised and under tension. The furled headsail is always raised, and therefore the halyard will be under tension. Ideally, you should relax the halyard when the boat is not in service, such as when you return to your berth after a weekend of sailing, as this reduces the strain on the rigging and sails. Sailors often forget to do this as their thoughts about the headsail disappear as quickly as the sail furls away. When they deploy their headsail again next time, the halyard might have creeped a little and the luff will be a little baggy. As discussed earlier, luff tension is important in positioning the location of the draft, so having a loose luff will make the draft move farther aft. After a few years of being in service, the halyard tension will all be gone, and the sail is going to be rather baggy. The simple remedy for this issue is to simply loosen the halyard a little to take the strain off of the headsail when you pull into your slip or anchorage, and tighten it again when you go to leave again. If you do this every time, you will always have the correct halyard tension for the conditions you are sailing in because you will be in the mindset of checking the halyard tension and therefore setting it accordingly to the conditions at hand.

DOWNHAUL

In the past, no sail could be safely flown without also having a downhaul attached. Now the mere mention of the name will make most sailors scratch their heads as sailboats are made without them. A downhaul is a line that attaches to the head of the sail to assist in bringing it down to the deck. This line is incredibly important to rig on all your non-furling sails, as it would get wrapped up in a furling sail and make a mess of things.

Under normal conditions, the halyard raises the sail and gravity lowers it. In storm conditions, the halyard raises the sail and the wind keeps it up there! If you are trying to lower the sail to reef or otherwise, without a downhaul you will find yourself struggling to grab onto any part of the sailcloth to pull it down as the wind keeps ripping it from your fingers.

The downhaul pulls the sail down from the top, the top of the sail comes out of the wind and the weight of it will make the rest of the sail finally drop down. For the mainsail, the downhaul is merely tied to the head of the sail and the process is finished. The bitter end of the downhaul should be made fast to a cleat near the tack

HOW RUNNING RIGGING TIES IN

of the sail with no slackness present; you don't want the downhaul to get fouled in the rigging and then work against you, holding the sail up and not letting it come down.

For hank-on headsails, having a simple turning block at the tack of the sail will do wonders, as this will allow you to direct your force parallel to the luff and stay. Hank-on headsails have a bad reputation for not wanting to come down in a breeze and flapping against your face as you try to wrestle them down onto the deck. This notoriety was part of the rise and success of roller-furling headsails, as all the problems of dealing with the headsail were remedied.

The truth is, if you have a downhaul tied to the head of the headsail, and a turning block at the tack, you can simply release the halyard and then pull down on the downhaul, bringing the headsail down and onto the deck with ease from a safe distance. Never again will you suffer the pain of being whipped by sheets or slapped by a sail because you have a downhaul!

CHAPTER SIX
How Sails Make It All Work

WHILE SAILS ALL WORK BASED ON THE SAME PRINCIPLE HARNESSING THE WIND TO propel the boat through the water, each type of sail falls into a few different categories based on where and when they are flown. For example, a mainsail is set behind the mast, normally on a boom and can be used for any point of sail while a symmetrical spinnaker is set forward of the mast and only used for sailing downwind.

Sails set behind the mast are located more toward the aft half of the vessel's center of lateral resistance (CLR), or balance point of the underwater profile of the boat. This point is the geometric center of the underwater profile, which means that the size and position of the keel and rudder will have a huge impact on the location of this point. If the keel and rudder were to move forward on a design, the CLR would move forward as well. If they are moved aft, the point will also move aft. The location of this point is fixed on most boats, unless you have a swing keel sailboat!

Boats with swing keels can lift their keel up to reduce their draft for getting into shallow areas, but can also lower their keel to improve their windward performance. As the keel swings from "raised" to "lowered," the CLR moves along with it, usually moving aft as the keel makes its journey to the "raised" position. If you have a swing keel sailboat, you can use the keel to help balance the sails, but for everyone else, this point is fixed and non-adjustable.

The best way to think about sail balance and the CLR is to imagine a seesaw. The fulcrum is the CLR, and the sails aft of the CLR will push the stern down and the bow up. The opposite is also true, where sails set forward of the CLR will push the bow to leeward, which will in turn push the stern to windward. This means that sails set forward of the CLR will cause lee helm while sails set aft of the CLR will cause weather helm.

How much force is put on the seesaw and where you are pushing on the seesaw is the geometric center of each sail. The mainsail pushes at its geometric center, also known as its center of effort, or CE, and the headsail pushes at its geometric center, or its own CE. The two CE forces then add up and if they balance out to be over the CLR point, then the seesaw will remain flat. If one of the two pushes harder (think: bigger sail) or pushes farther from the fulcrum (think: long bowsprit or a yawl's mizzen), it will then have more of an effect on the seesaw and in turn the balance of the boat.

On a swing keel sailboat, the fulcrum point can be moved to adjust and compensate for the sails combined CE, but for the rest of us, we need to make these adjustments to the sails themselves as they are the only part that we have control over. This is why flying only one sail, mainsail or headsail, yields a slow and unbalanced sailboat where you have to overuse the rudder to fight the way the wind blows your boat.

Naval architects and designers can position the CLR wherever they deem fit to balance out the sail plan above. This is why having a sloop with a mast set far forward would not be plagued with lee helm, as the keel and rudder are positioned to accommodate and balance out the sail plan above. A cutter, by the same regard, with its mast set farther aft is properly set and balanced to the keel below, also set farther aft to make the CLR and CE line up perfectly, resulting in a well-balanced yacht that sails in any direction.

In this chapter, we will first discuss the different types of sails and how they work individually, then as a whole, and then we will look at how rigging plays a role in their performance.

SAILS SET BEHIND THE MAST

Sails that are set behind the mast are called by the mast they are flown from. These would be the mainsail, mizzen sail, and foresail. The mainsail is the most common as every boat with a mast will have one. On boats with multiple masts, mizzen sails and foresails can exist because these boats—ketches, yawls, schooners, or any other multi-masted sailing vessel—will have the additional masts to fly the additional sails. For every sloop or cutter floating in a marina, they will only have one mast and that mast is the mainmast, flying the mainsail.

These sails are typically set with a boom to grant them greater trim control. Picture a large headsail eased to leeward, the sheet is loose and the sail is twisted in a way that causes it to spill its wind, losing a lot of potential power. If that sail were poled out, suddenly it has a lot more potential to be trimmed for better efficiency. The pole that runs along the foot of the headsail is precisely what the boom is; just a permanently rigged pole on the boat.

These sails use a boom to grant you incredible control over the foot of the sail, no matter the position the sail is set. If the sail is sheeted over the centerline, or eased all the way out, perpendicular to the boat, the foot is always under control thanks to the outhaul.

The leech is controlled by the mainsheet when it is sheeted in, and the vang when it is eased out. You might be thinking: "I thought the mainsheet was supposed to be used to trim the mainsail and not just to control the twist of the leech?" This is a common misconception as the mainsail is actually trimmed by the traveler, while the mainsheet is used to control twist. There is a point when the traveler simply ends and the boom needs to go farther to leeward, which is when the mainsheet takes over and the twist control then falls to the vang. Once again, the advantage of a wide beam demonstrates itself in the form of a long traveler track for greater mainsail control on a wider range of points of sail.

This is all well and good in fair weather, but during a storm, the boom becomes a massive liability! The greatest mental stress to the crew is the fear of the boom hitting something or someone and causing a great deal of damage or death. Removing the boom from the equation would make everything easier and that is where the trysail comes into play.

The trysail is one of the few sails set aft of the mast that doesn't employ the boom. This sail looks like a small headsail, and is flown in lieu of the mainsail during storm conditions. Its size is similar to a mainsail with three reefs set, and it is made out of incredibly heavy sailcloth. This is a specialty sail only flown in very specific conditions.

With it being the size of a third reef, you might be wondering, "Why even bother rigging up a different sail when the main is already there?" The reasons are twofold: First, you do not risk damage to the head of your mainsail, and second, you remove the boom from the equation.

The mainsail is made out of sailcloth designed to be good

enough in light air conditions but also strong enough to handle heavy wind. The cut of the sail is also designed for the general purpose of sailing in these varied wind speeds. When you reef down, you will be using a small portion of a sail that works well in light winds and is made out of a general-purpose weight of cloth. When you compare that to a sail that is cut to work well in storm force winds and made out of a sailcloth specifically selected for this purpose, the choice is obvious why you should use a trysail when sailing in a storm that could end your life if things go sideways. If your sail blows out and you lose control of the boat in a severe storm, you are screwed! The reason for this book is so you can learn to properly set up and maintain your boat with the right gear, making these storms just something you have to wait out as they blow over because you have assembled a rock-solid rig. Why would you then hang a small portion of a thin sail in that breeze?

Second, the trysail eliminates the boom from the equation. Various sailors have different methods for dealing with the boom. Some like to sheet it in centerline, others lower it to the deck and lash it down, others like to use the preventer to swing the boom far to the side of the boat. In any of these situations, the boom is immobilized so that it can't swing around and hurt anyone or anything. Jibing the trysail is just as easily accomplished as jibing a headsail; all you do is change course and when the sail flops over, switch the sheets to the other side!

As for the three methods of stowing the boom in a storm, my personal favorite is to prevent it out to the side. My reasoning for this is as follows: Pulling the boom centerline means that the trysail sheets need to be routed over the boom but forward of the topping lift. If you forget to do this, then the sheets will foul the boom or the topping lift and make a mess. Having the boom centerline is the most straightforward. All you need to do to get it there is crank in on the mainsheet, but then you need to carefully feed the sheets through a variety of places on a boat that won't be the most stationary as the winds build.

Lowering the boom onto the deck sounds like a better plan, as it gets it out of the way of the trysail and also holds it steady so it can't shimmy around as the sheets stretch during the storm. The only reason I don't like this approach is it then obstructs that side of the deck. If you lowered the boom onto the lee deck, your windward deck is clear if you need to go forward to the mast or bow for any reason. If you tack, now the windward deck is blocked and you have to climb out on the low side of the cockpit. It is a good practice to always walk on the windward deck, as this is the high side; if you fall, you fall on the boat, but if you are on the low side and fall, you are going overboard.

Preventing the boom out to one side gets it out of the way so that the entire deck is opened up, and frees up the entire area for the trysail rigging. Which side you prevent the boom out towards depends on which side of the mast your trysail track is placed. While the trysail can be rigged into the mainsail luff groove, this is not a good practice as you want to be able to switch to the trysail with as little effort as possible. This means that the trysail should have its own track and halyard. Personally, we never leave port or anchor without bending the trysail onto its track and attaching its halyard. If the situation should arise, all we need to do is toss the sheets into the cockpit, prevent the boom to leeward and raise the trysail. In our case, the trysail track is on the starboard side of the mainsail track, so we prevent the boom to port. If your track were on the port side, you would want to prevent your boom to starboard.

The sheets of the trysail should be led aft in the same manner that you would for a jib, where the sheet angle approaches the 40% mark on the luff. To make this easier to accomplish, the clew is normally set lower than the tack so that the angle is easily achieved. The sheet blocks should be dedicated for the task at hand that way the sheets can be quickly rigged up at a moment's notice.

The standing rigging plays a major role in controlling these sails that are attached to the mast. First of all, the mast shape will directly affect the luff of these sails as the luff is attached to the mast. If the mast is straight as an arrow, so will the luff, the same holds true if the mast bends or bows fore-aft and side to side. Your sailmaker has accounted for the shape that your mast is supposed to be, having a slight bow backwards along with a slight rake aft.

To keep the mast in column as the boat heels over, you need to make sure that the shrouds are all adjusted to the appropriate tension. This is important to make sure the luff of the sail remains straight so that it can work as designed. Backstay tension is very important for the shape of the mainsail because the backstay will cause the mast to rake aft, which will cause the middle portion of the mast to bow forward. This action will flatten the mainsail and cause the angle of attack to decrease. The result is a flatter mainsail, which will point higher; it will also depower the mainsail, which is useful if you are trying to delay reefing because conditions look like they are about to improve.

The opposite is also true: Easing the backstay will cause the masthead to move forward slightly, which will make the mast straighter and give the forward third of the chord of the sail more sailcloth. This will result in a deeper draft and more power in the sail. Having the masthead move slightly forward will also cause it to generate a little more lee helm, which is advantageous when reaching or running.

With a trysail, the rigging is less of a concern as the sail will be kept lower on the spar. Having the head of the trysail at the height of the first spreader will ensure that the forces on the rig will be directed to the lower stays, which have the easiest time doing their job due to the angle that they pull from. These stays do not need spreaders to achieve the minimum angle necessary to accomplish their task, which is why these stays are also the most effective. Keeping the loads down to these stays makes the entire storm situation easier on your rig.

HEADSAILS
Jib

In the world of headsails, you can safely break it down into upwind and downwind sails. Some of these sails can be used for more varied purposes while others are extremely specific. The most common headsail and the one with the broadest definition is a jib.

> Jib: *n.* triangular staysail set forward of the mast.

Excellent, that pretty much sums up every single sail on a modern production boat. Looking more closely at the definition, we see a couple of key terms. The sail needs to be triangular, meaning it has the three standard points of attachment—head, tack, and clew—and it is stayed. Size and shape are not mentioned in this very basic definition because that opens up an entire chapter of discussion and details.

To sailors, the term "jib" refers to a specific type of headsail, one that is stayed and set entirely inside the foretriangle of the boat. The size of the jib is also listed as a percentage, alluding to the length of the foot relative to the distance between the tack and the mast, also known as the J measurement on the IPEJ system.

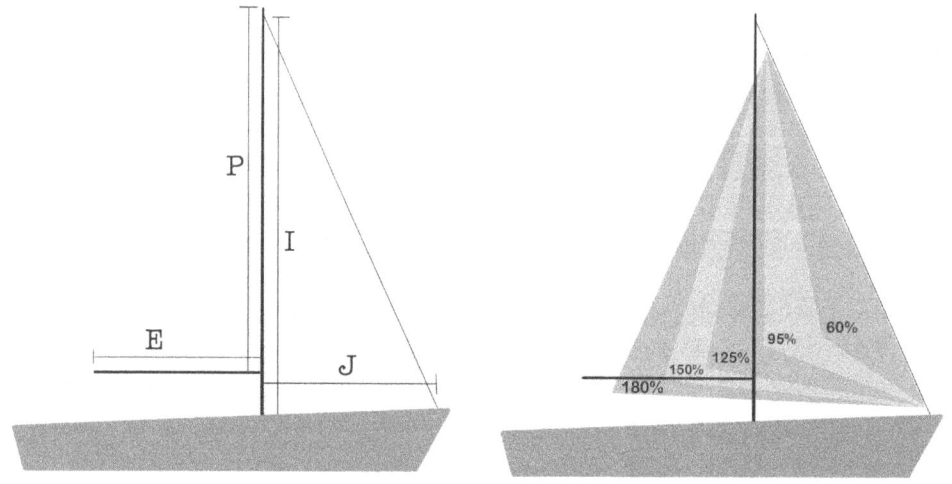

The largest jib could be a 100% jib, and this would place the clew at the front of the mast. This size is typically not done because it risks fouling of the clew or leech on winches or other gear on the mast. Jibs tend to top out at the 95% jib size.

The percentage rating of a sail is not a good comparable for sail area between different boats. The percentage is relative to that boat's J measurement, not looking at the actual length for the J measurement or paying any respect to mast height. Imagine you have two sailboats: *Ariel* with a J of 10 feet with a mast height of 30 feet, and *Windsong* with a J of 20 feet with a mast height of 50 feet. *Ariel* has a 95% jib and *Windsong* has a 40% jib. Which is bigger?

Ariel: $(10 \times 0.95 \times 30) \div 2 = 142.5$ *square feet*
Windsong: $(20 \times 0.40 \times 50) \div 2 = 200$ *square feet*

That wasn't overtly obvious, hence why we need to convert the sail size from percentage to area to properly compare between sails on different boats.

The nomenclature for sail sizes on a boat is relative to the other sails onboard. You might have heard someone mention when they fly their J1 versus their J3. This statement will tell you two important things: First, you are talking to someone who likes to collect sails and has found a way and manner to store all of these sails in their boat! Second, they have the same style of jib, but cut into various sizes for different wind conditions. The sails are always numbered, starting with 1 and becoming smaller as the number grows. If they have five different jibs, they would be numbered J1, J2, J3, J4, and J5. The J1 would be flown in the lightest of wind conditions while the J5 would be for the heaviest of wind conditions. As with percentages, these numbers have no bearing on the relative size of the sails on another boat and must be converted to units of area to properly compare sail size.

Some common traits of all jib sails are that they are used predominantly for upwind sailing and they have a low clew to give absolute control of the sail when sheeted in tightly. Having a low clew, the tension of the sheet is transmitted into the leech of the sail when the car is far forward, allowing the sail to become a powerful aerofoil to generate a lot of lift with minimal drag, thanks to its high aspect ratio. These kinds of jibs are sometimes referred to as a blade jib when they have a flat draft making their angle of attack very low. They can create a lot of power, thanks to their long luff, while also pointing incredibly high upwind.

The biggest factor affecting the performance of these sails will be headstay tension. If the headstay is super tight, the luff will remain straight and the sail will be flatter. As a result, the sail will also perform better to windward in strong winds. The opposite is also true, a looser headstay will lead to more headstay sag, which will cause the luff to fall to leeward and this will make the draft of the sail greater. A greater draft will produce a lot more power but the angle of attack will suffer and the sail won't be able to point as high to windward anymore.

Headstay sag is controlled via the backstay tension. As the backstay is tightened, the masthead is pulled back and that exerts a force on the stay that is opposing this action, which pulls on the headstay. The position of the headstay relative to the masthead has a further effect on the amount of tension the backstay can generate, and this will be discussed further in chapter 7.

Genoa

As soon as a jib exceeds the 100% size, it then overlaps the mast and becomes a genoa. Once again, these sails are named based on their percentage to their J measurement and not to any unit of area. In our example between *Ariel* and *Windsong*, if Ariel has a 150% genoa and *Windsong* has a 105% genoa, we need to do calculations to figure out which sail is bigger and how long the foot of each sail would actually be.

Genoa sails are very common on sloops as the mast is so far forward that the foretriangle is extremely small. The resulting jib would be ineffective and as a result would lead to a sailboat that is difficult to power. By extending the headsail aft of the mast, the headsail becomes a size that is respectable and can then function in conjunction with the mainsail to power the sailboat in various wind conditions.

If a boat has multiple genoas, they would be numbered using the same system, but using the letter *G* instead of *J*: G1, G2, G3, and so on. On most modern boats, the need for multiple genoas has been removed because of roller furling, which lets you have one large genoa and via roller-furling reefing, you can make it smaller. This gives you the advantage of having many different sizes of headsails while never needing to change the headsail or finding places to stow all of the headsails.

Aside from the increase in sail area, genoa sails offer advantages that a regular jib cannot. The first major difference is the location of the center of effort. The CE of a jib will always be forward of the mast, as the entire sail is located up front. This means that the CE from the jib will always induce lee helm. A larger jib will induce more lee helm than a smaller jib, but both will produce results that contribute to lee helm.

As a genoa moves aft of the mast, its CE moves aft as well. If the genoa is large enough, and the boats CLR far enough forward, the genoa can actually create weather helm, something that no jib could ever do! This can cause confusion for sailors who are having trouble with weather helm, so they get a bigger genoa to help give them more lee helm to balance the boat, and on their first test sail (after spending a lot of money on a new sail), their problem has been exacerbated!

The second feature that is unique to a genoa is the slot effect, caused by the increase in velocity of the air that is coming off of the genoa and onto the mainsail. The increase in velocity causes a decrease in pressure that makes the low pressure on the leeward side of the mainsail even more drastic, and as a result more powerful. This is the same method used by airplanes when taking off and landing, when they are flying at lower speeds and need to generate more lift. The wing opens up into a variety of sections that allows them to generate more lift. The slot between the genoa and mainsail operate via the same principle, where the power generated by the two sails is greater than the power each sail could generate independently.

This second aspect has rocketed sloops with their overlapping headsails to the front of the market for modern production boat designs. Regardless of the size or style of sailboat, there will most likely be a mainsail and a genoa on a furler as part of the standard equipment on the boat.

Yankee

A jib, with its low clew, is the ultimate sail for upwind performance and control. As you start to sail off the wind, the jib would need to be eased. The leech is significantly longer than the foot, so the forces on them are not even and will result in the leech twisting freely, causing it to spill wind and become less effective. Next time you see a sailboat with a genoa sailing downwind with the sail eased, notice how the top of the sail twists forward, spilling all of its wind.

In order to make these forces be more evenly distributed between the two sides of the sail, you will need to make the two sides of the sail the same. The result is a smaller sail with the clew set almost in the absolute middle of the sail. As this sail is eased, the forces are evenly distributed between the leech and the foot, which results in a sail that is eased for reaching with plenty of control. These sails are commonly found on ocean crossing cruisers as they are predominantly sailing downwind routes where a reaching sail is more valuable than a beating sail.

A yankee can overlap the mast but this is not commonly seen as the yankee normally is seen on the headstay of a cutter, with a staysail set aft to it. Yankee sails were also commonly seen on tall ships set in vast numbers as they made the sail area more manageable by breaking it up into smaller sails that could be controlled without the aid of winches.

A variation of the yankee jib is the flying jib, which as its name states, is not attached to a stay. This sail is set flying, held in place by merely the tension of the halyard and the tack line. The flying jib was normally a small Yankee jib set high above the other jibs to help fill in the foretriangle in lighter air conditions. Over time, this defining characteristic of being set flying has been ignored and the flying jib found its way onto a stay. The definition of a flying jib gradually changed to mean that it was "flying higher than the other jibs."

The flying jib has the special property that allows it to adjust the position of its CE by moving forward or aft along the stay. If the sail is set all the way at the top of its stay, then the CE moves aft and can help reduce lee helm. If the sailing vessel is on a reach where lee helm is desired to help maintain course, moving the flying jib down, and as a result forward, moves the CE forward. This sail helps both with positioning the sails, CE as well as grabbing speedier winds that blow at higher altitudes.

Staysail

By definition, a staysail is any sail that is flown attached to a stay, while a flying sail is any sail flown without being attached to a stay or spar. By this broad definition, every single jib and genoa on every sailboat in a marina would qualify by this definition as a staysail. Thankfully, the sailors who named everything thought of this confusing point and ignored the problem, carrying forward with calling this one particular sail a "staysail."

A staysail is any jib that is set on a stay aft of the frontmost headstay. Most of these will be on cutters, with the jib, yankee jib, or genoa on the headstay, and then the staysail on the inner forestay. The reason we can't say staysails are on the inner forestay is because there is a type of schooner called a staysail schooner, which is so named because the triangle where the foresail would fly is occupied by a staysail running from the top of the mainmast to the bottom of the foremast. Imagine a normal sloop with a 95% jib, now pop a second mast ahead of the tack and then follow up with some extra boat up front to support the headsails for the foremast. This is what a staysail schooner looks like, and they are gorgeous!

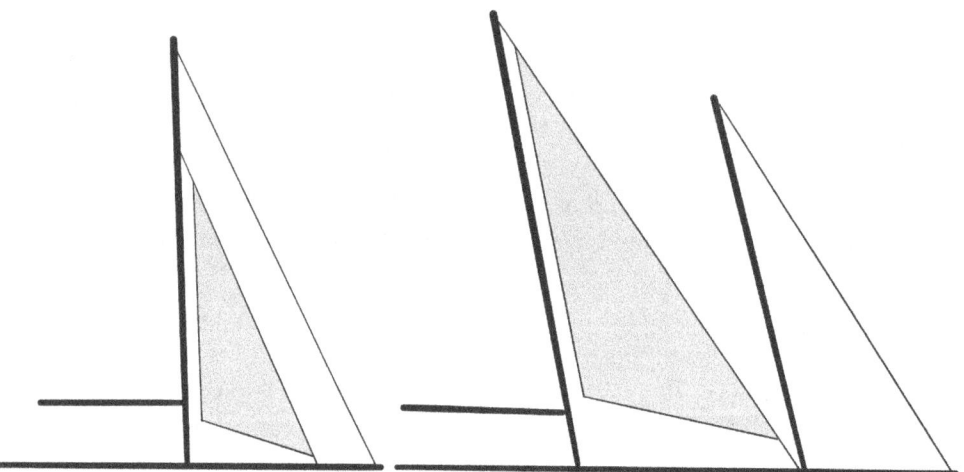

The staysail on a cutter is a special sail that accomplishes many tasks while only imparting a few drawbacks. The first thing to note about a staysail is the shape and position. It is seen predominantly on cutters; since these boats have their mast set farther aft, there is automatically more deck space ahead of the mast to fit more sails. If a sloop needs to have an overlapping genoa to create enough sail area to be effective, then there is no way that a tiny sail occupying only a portion of the foretriangle would be effective at harnessing the wind.

That said, there are genoas that are set on the inner forestay and they are called staysail genoas. The issue with this sail is that tacking becomes very stressful, as the staysail's clew will drag past the mast, likely snagging and fouling on something during the process. The second issue is this is not going to be the only sail flying on the sailboat, so tacking becomes even more stressful. For this reason, staysail clews are often kept ahead of the mast; that way tacking is easier and it even opens the door for self-tacking staysails. These systems use a variety of options, ranging from booms, tracks, or bridles that allow the staysail to tack back and forth on its own, which makes short tacking much easier, as you have one less sail to work.

The other reason that staysails are kept ahead of the mast is they are also cut in a shape that is similar to a storm jib. This means that the staysail is also the storm jib, freeing up a spot in the sail locker, and reducing the hassle of switching the headsail over to the storm jib as the wind starts to pipe up.

The staysail is technically just a smaller jib, so you might be wondering why bother having this sail when you could just reef your jib down to a smaller size? The answer has a lot in common with the trysail, first, where the sailcloth is chosen for the strict purpose of handling high winds. The second has to do with the position of this smaller sail.

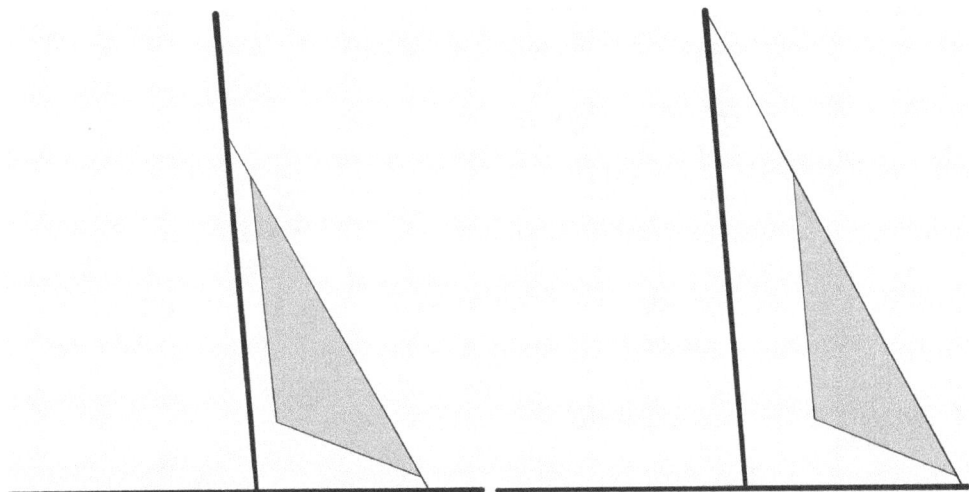

As a sailboat reefs its headsail with roller furling, the CE actually moves forward and higher; with regular reef points, the CE moves forward and lower. As you reef your sails, you want to make them both smaller, but also closer to the middle of the boat. This makes the boat easier to control, as the CE of each sail and the combined CE of all the sails, coming closer to the point where the underwater CLR resides. By moving the sail closer to the middle, it also reduces the lever arm of the sail to pivot the boat.

A tiny jib set far forward at the tip of the bow will be able to turn the boat to leeward in a gust a lot easier than the same sail set right in front of the mast. This means that the rudder will have less work to do and can focus on steering the boat instead of fighting the sails. Gusts in a storm are very common, and as the wind builds, so will the seas; when the sailboat is down in the trough between waves, the windward wave will block some of the wind, making the conditions feel slightly less gusty. When you rise up to the crest of the wave, you have sailed out of the windshadow and will be hit by a gust of the storm's breeze. As you surf down the wave, you will fall back into the windshadow of the next wave. This horrible process will repeat itself for hours, even days, as you ride out the storm. If your storm jib is far forward, every wave crest will be accompanied by a leeward push on the bow that the rudder has to fight. If the sails are set closer to the mast, the helmsman will have a less fatiguing job to do as the waves will simply bring a gust of wind for a few seconds as the boat continues to sail straight on its course.

While staysails are but tiny jibs set inside the foretriangle, they are, in my opinion, the most useful sail on a sailboat. They are heavily constructed and easy to manage due to their small size relative to the other sails on the boat. They also find ways to be incredibly useful in all situations. If you are on a run and the wind is fluky, you can have this be the only sail flying on a self-tacking bridle, allowing it to flop from tack to tack as the wind causes you to jibe at random times while sailing on a straight course. When raising anchor single-handed and under sail, having this sail up with the rudder turned hard to one side will make your life much easier. The staysail will sail you up onto the anchor, making the prospect of cranking in each link of the chain much less daunting, and once the anchor is free, the boat will sail in a circle while you race to get the anchor from the seabed to your bow roller. The staysail will simply drive the boat forward while the rudder has you turning in a constant circle. There is no risk of damage or uncontrolled jibing as the staysail will simply flop around as the boat changes angle to the wind. This will give you a moment to get the boat situated before getting on a course and raising sails.

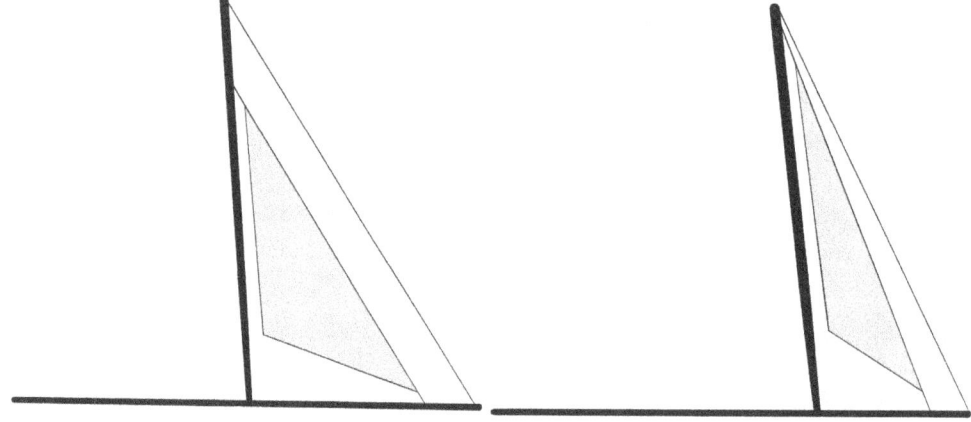

If you have a boat with a staysail, you will find that this is the first sail you will raise and the last sail you will lower. The staysail can be flown on all points of sail and in all wind conditions. In light wind, it will give you a little more sail area for added drive, while in heavy winds, it will become part of your storm suit of sails. For upwind sailing, it can be sheeted to an inboard track, while on a reach it can be sheeted to the toerail with the aid of a barberhaul. This sail is so versatile that cruisers will often try to add them to their sloops, turning them into a rig colloquially termed a slutter rig.

Spinnaker
When it comes to downwind sailing, there is no sail better suited for the job than a spinnaker. The spinnaker sail is a massive sail that will pull a sailboat along as fast as humanly possible. Spinnakers are often flown in light airs where the ultralight nylon cloth can catch every last bit of breeze. While these sails are not very strong and shouldn't be flown in high wind conditions, we have all seen sailboats flying these in races as they exceed hull speed with whitecap waves all around them. The reason the spinnaker sail survives this torture is because the apparent wind on the sail is not that great. If the true wind is blowing at 20 knots, but the boat is surfing down waves at 15 knots, the apparent wind on the spinnaker would only be 5 knots!

These sails work more on the principle of generating drag than they do on generating lift, which is why they are only used for downwind sailing where a lot of drag set far forward will pull the boat along on a downwind course.

Since this sail only operates in downwind conditions, it will always be exposed to a lower apparent wind than the true wind is. For this reason, these sails need to be as big as physically possible, often equaling if not surpassing the total sail area of all the boat's working sails. To allow even bigger sails than what would fit on the boat, these sails are often poled out well beyond the bow, allowing for even more power to be generated by this colossal sail area.

The symmetrical spinnaker is a symmetrical sail, and is the classic spinnaker that every landlubber envisions when they think of a sailboat. This sail pulls the boat dead downwind on a run and can reach to a minimal degree. To improve the reaching ability of these sails, asymmetric spinnakers have been developed.

Asymmetric spinnakers, or asymmetricals, are cut with three distinct corners and range in size depending on the desired point of sail for it to be used on. Asymmetricals being flown on more points of sail has led to the need for a better way to deploy and retrieve the sails, and this has led to a wonderful growth in technology on sailboats.

Traditionally, spinnakers were set and retrieved by a skilled crew who were both trained and numerous. This meant that the crew in the cockpit could focus on their job while the deck crew would raise or lower the sail, managing it along the way to

prevent it from twisting or fouling. As asymmetric spinnakers became more popular, the idea of setting and striking these sails on a reach added to the complexity and catastrophe of the situation. These sails also started to become popular on cruising yachts, usually crewed by a couple who now simply do not have enough deck crew to wrestle with the spinnaker as it is being raised or lowered. The first method to simplify spinnaker sailing is called "the sock."

The sock simply pinches and bags the sail inside a sock, removing it from the wind and making the sail turn into a very long worm in the air. This worm is raised or lowered with ease and the sock is lifted to the head of the sail to set the sail, or pulled down over the sail to strike it, making this massive sail very manageable.

The sock works great while it's working great, but if it snags on anything or simply gets stuck, you now have a full-size spinnaker to contend with without the practice. As technology continued to evolve, furling systems came onto the market. The favorite for spinnakers is the top down furler, which simply winds the head of the sail around a torsion rope and pulls the rest of it into the bundle as it spins. This system works really well, but it requires everything to be set up and maintained perfectly. If any point is overlooked, it will quickly fail and your life will very quickly become complicated.

These sails have revolutionized downwind sailing for the average cruiser as they are now able to sail better in lighter winds and put the massive sail away quickly when the conditions change. While these sails are amazing, they also need a healthy amount of respect. I have seen spinnakers get out of control and wrap around the rigging, getting fouled in a way that they cannot be lowered without cutting the sail free. If these sails get out of hand, they have enough sail area to pull the boat all the way over and when they fill with water, they are even harder to handle. As with everything on a sailboat, they work great while they work great, but the forces on them are tremendous and they should be respected. At your first thought about bringing the sail down, you should begin doing so! These large sails are always easier to handle in calm airs instead of waiting for the storm to strike to begin bringing the sail in.

Special Sails
The sails discussed above are sails that you will find on most boats and definitely sails that you will hear talked about at the marina tiki bar. If you want to wow some of the salty old sea dogs at the bar, you can always mention some of these specialty sails. They are specialty sails because they are only used for very specific situations or flown on very specific boats. As a result, they carry a bit of rarity to their name.

The first two are the gollywobbler and the fisherman. These sails are set between the two masts on a schooner. Since schooners are incredibly rare, these sails are as well. The gollywobbler, aside from having the best name of any sail on a boat, is a light air sail. This sail will fully fill the entire space between the foremast and mainmast, adding a ton of sail area and allowing the schooner to power its way on the lightest of zephyrs. The fisherman is the heavier wind version of the gollywobbler, flown at the top to fill in any voids in the sail plan between the foremast and mainmast. The

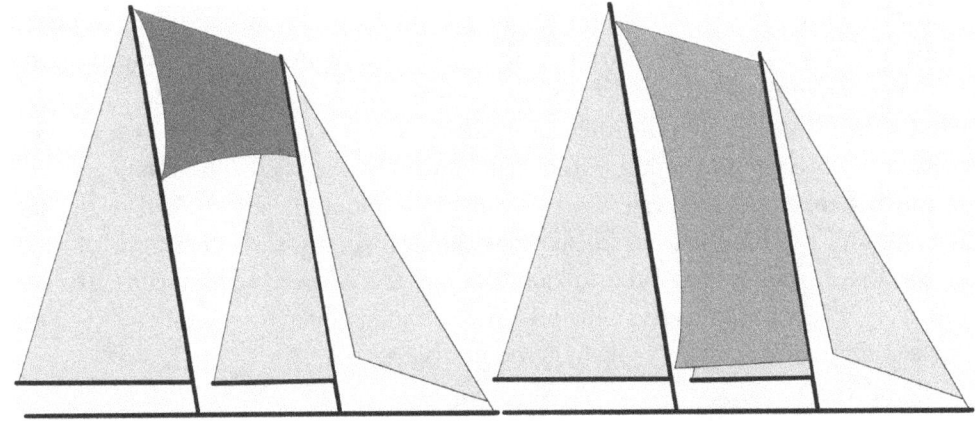

fisherman is much smaller and therefore won't generate as much power as the gollywobbler, but it will take advantage of the altitude to catch the fresher breeze that always blows up there.

Moving on to ketches and yawls, we have some sails that are flown from the front of the mizzenmast. The first is a staysail and the second is a spinnaker, aptly named mizzen staysail and mizzen spinnaker. As you can imagine, these sails would be in the way of the mainboom during a tack, so these sails need to be doused and reset every time the boat tacks. That is a lot of work, which is why these sails are only set when the boat is on a long course and won't be tacking anytime soon.

The next specialty sails are known as studding sails, and they are light air sails that are flown next to the edges of a square sail. Studding sails add sail area to square riggers and gaffers by making the sail bigger to the wind. In a time before diesel engines, when ships relied on the wind for propulsion, studding sails were a valuable piece of equipment aboard as they meant the difference between sailing to the next port and bobbing around waiting for wind.

The next specialty sail works along the principle of studding sails, where it simply adds sail area to the boat in light winds to try to drive it even faster. This sail is the blooper sail, which may have fallen out of favor just because of its name. The blooper is a second spinnaker that was flown under and around the spinnaker. It was basically trying to catch any extra wind available to give the boat that added oomph to win the race. These sails were set opposite of the spinnaker and were very challenging to control and keep full. As spinnaker design improved, the supposed advantage of having this sail quickly outweighed the value of having it on board. That said, next time you see an old sailboat race photo in the dockmasters office where the sailboats all have two spinnakers set, you can now point to the smaller sail of the two and tell the dockmaster that "that sail there is called a blooper."

The drifter sail is a bit of a misnomer as this giant light-air sail will keep you sailing instead of drifting. Drifters are massive sails cut like a genoa with a bit more belly in them to give you more power in light airs. They are made out of spinnaker cloth, also known as rip-stop nylon, making it a giant spinnaker on a stay. By virtue of being stayed, it can be used for sailing upwind as well as downwind in light air conditions. More importantly, the sail is attached to a stay so when it is lowered, it is easier to manage as the luff will always remain at the stay. While these sails are normally hanked on, they can also be set flying without any attachment to the stay, but this will preclude you from beating as the luff will sag too much to be effective to windward. The decline of this sail almost mirrors the rise of roller furling. The ease of having a larger than usual genoa on a furler that can be partially rolled out to suit the conditions is far easier than changing headsails.

THE PURPOSES OF DIFFERENT SAILS

Working sails are the sails that your boat will fly in most weather conditions. These range from light winds to moderately heavy winds. If the wind conditions are lighter, you will want to fly light air sails; the same holds true in storm conditions where you will want to fly storm sails. Upwind sails and downwind sails are equally important to have and deploy at the right time. Upwind sails are flatter and work by generating more lift than drag, while downwind sails have a deeper draft and work by generating more drag than lift.

Each sail needs to have its own rigging set precisely for it and the rigging needs to match the needs of the sail. For light air sails, it is important that the rigging be exceptionally lightweight so that it doesn't weigh down the sail and cause it to lose its shape. Storm sails need to worry about the opposite problem, this rigging needs to be very strong as it will be subjected to intense loads and the survival of the ship is riding on this rigging.

One important point with any sail is that when it is set, it is also still. If the sail is flopping around, it is not properly set and will cause damage to the sail and rigging. The most damaging motion a sail can do is to flutter. This rapid oscillation will harden the fibers of the sail and cause them to become brittle and break as well as rattling out the resin that holds the sail cloth together. Trimming the sails properly with the running and standing rigging will ensure that the sail as well as the rigging lasts as long as possible.

Reefing is a valuable topic that merits mention again here. Making the sail smaller not only reduces the load on the sail, but also reduces the load on the rigging. This will make the boat sail more upright allowing the keel to better counteract the leeward force of the sails, which will make the boat sail better overall. A sailboat is designed to sail at a particular angle, usually less than 30° of heel. When the boat heels more than that, the sails become less effective as they are laying on their side instead of facing the wind, and the keel loses effectiveness as it is not sufficiently in the water. The result is the boat will slip to leeward as the sails flap against the surface of the water. Reefing brings everything back into equilibrium and allows you to sail on the boat's designed lines.

An old adage is "reef early and shake late," meaning put the reef in the sails anytime you think you might need to and if you are thinking of shaking out the reef, give it a few minutes to make sure that the winds aren't about to return. This mantra will keep your boat, your sails, and your rigging under control and in good working order for a long time.

CHAPTER SEVEN
GOING TO THE TOP

Where the headstay attaches plays a big role in upwind performance, but also detracts from the ultimate amount of sail area that can be flown. The balancing act between these two factors can lead to an ultimate upwind racing machine or a relaxing cruising yacht.

When the headstay runs all the way to the top of the mast, this is known as a masthead rig. If the headstay doesn't make it all the way to the top, this is known as a fractional rig. Having the headstay run to the tippy top of the mast means that the headsail can fill the entire foretriangle of the boat, extending from the tip of the bow to the top of the mast. If the headstay were set any farther aft of the stem or any lower than the masthead, then the total foretriangle area will be reduced and so will the sail area.

Fractional rigs are named based on the fraction that the headstay meets the mast. If headstay meets the mast three-quarters of the way up, it will be referred to as a three-quarters fractional rig. Another common style is the seven-eighths fractional rig where the headstay meets at the seven-eighths mark and only one-eighth of the mast extends above the headstay.

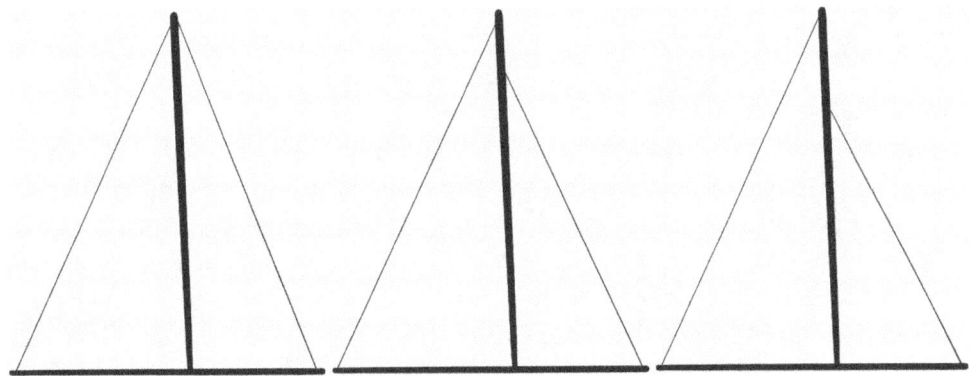

The reason that fractional rigs are preferred in racing applications is because the mast will be even easier to adjust for optimal sail shape. A tighter headstay will have less headstay sag and allow the headsail to point higher upwind. Having a slight bow in the mast will also flatten the mainsail and allow it to point higher upwind. Naturally, all of this needs to be done on a mast that is bendier, which is why these masts tend to have multiple sets of spreaders and maybe also be deck stepped, as discussed in chapter 4. Having a bendier mast allows you to take advantage of the mechanical advantage that is generated by not having the backstay transition into the forestay at the masthead.

The lever arm created by the top one-quarter or one-eighth of the mast allows the force of the backstay to be magnified against the headstay, allowing you to make it even tighter to boost windward performance. This lever arm also grants the backstay improved ability to bend the mast forward, giving you better control to flatten the mainsail. This allows for ultimate sail shape control making the sailboat sail to windward with ease, at the cost of sacrificing some sail area of the foretriangle.

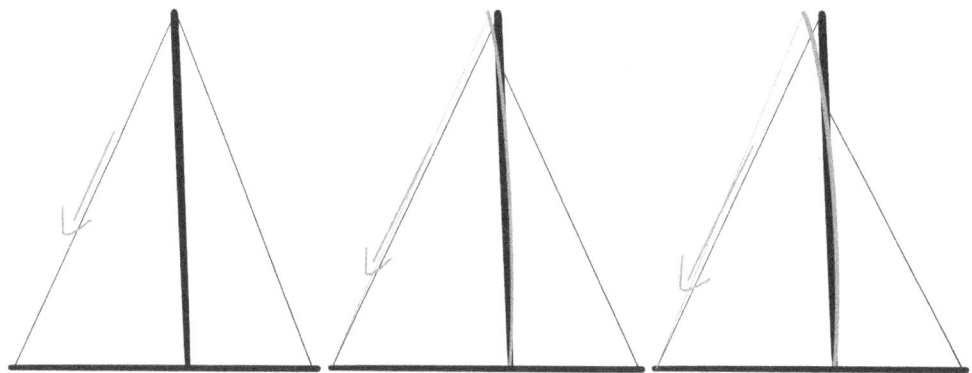

A smaller foretriangle is not that big of an issue when the foretriangle doesn't dictate the maximum size of the headsail. Setting a genoa on the fractional headstay grants the sailboat the ultimate combination for upwind performance. Perfect trim for the mainsail, ultimate tension for the headsail, and the slot effect to create the synergistic increase in performance of the sails.

If you know sailors, you know that racers and cruisers blend together like oil and water. Racers want a bendy mast that they can tune to perfection and squeak out that last bit of performance for the win. Cruisers want to get somewhere without doing a lot of work. Imagine a cruiser trying to eat their lunch while their dog is barking at a pelican and negotiating with tugboats off the coast between one anchorage and another; now imagine that same cruiser with a bendy mast that needs constant adjustment! That cruiser would probably say, "I have too much else going on; drop the sails and turn on the diesel because I need to make it there by sundowner time."

Cruisers want safe, reliable, and robust, which is exactly why they tend to prefer boats with masthead rigs. For a masthead rig, backstay tension doesn't do much to bend the mast to improve the shape of the mainsail, but it will tighten the headstay to reduce headstay sag and that in itself will greatly enhance the sailboat's windward performance. Cruisers don't really seem to care about the mast shape and how it affects the trim of their mainsail because their mainsail is probably getting a little baggy after many years of constant service anyway. They might not even have a way to adjust the tension of their backstay on the fly; they just want to enjoy the sail and not work hard at it, so having a mast that is stout frees their mind from concern.

If you already have a sailboat, enjoy it. If you like to race and have a masthead rig, fly the biggest headsail you can and crank on the backstay to tighten the luff so you can slice through the water on your windward course. When you turn to come back downwind, easing the backstay will ease the headstay and cause that giant genoa to billow and fill with air, pulling you downwind with plenty of lee helm to make steering effortless.

If you are a cruiser and have a fractional rig, you can find a sweet spot where the sails work well enough and then leave things alone so that you don't have to mess with tweaking the mast while trying to eat lunch as the family and pets roam around the boat entertaining themselves. One rig style does not preclude you from the other activity, but if you don't have a boat yet and are more interested in racing or cruising, then this chapter can help you narrow your search based on rig styles.

CHAPTER EIGHT
DOWN TO THE BOTTOM

MOST PEOPLE LOOK UP THE MAST WHEN SOMEONE ASKS THEM TO LOOK AT THE rigging. This is a logical thing to do as most of the rigging is located overhead. The most critical part of the rigging though is located at the height of your toes: the chainplates.

Chainplates are merely bars of metal that transmit the load from the rigging to the hull. They are attached to strong members of the boat that are designed to receive these intense loads and transmit them throughout the rest of the hull. The reason that most chainplates are bars of metal is simple: This is an easy way to connect the rigging and to the boat.

At the top of a chainplate, there will always be a hole for the clevis pin to attach. This is an incredibly strong junction as the hole is perfectly sized for that specific clevis pin so that the entire load is evenly spread out over the large metal body of the pin. The size of the hole in the chainplate depends on the size of the clevis pin, which is determined based on the size of the standing rigging that connects to the mast.

Below, the chainplate is attached to the boat in one of two fashions: internal or external chainplates. External chainplates are the old way of doing it where the chainplate wraps around the outside of the topsides and is bolted to the hull and frames. The hull needs to be reinforced in this area to sustain the additional load that will be imparted upon it. Internal chainplates provide a more aesthetic result with the chainplate slipping through a hole in the deck and attaching to a structural member inside the boat, normally a bulkhead.

In both of these situations, where the chainplate meets the boat needs to be incredibly strong as all the force of the wind pushing the sailboat over will be transmitted to the hull via this junction. The chainplates are pulling the hull up as it lifts the ballast of the keel. At some point, the wind pushing on the sails is countered by the weight of the ballast in the keel and the sailboat reaches equilibrium, sailing along with a designated angle of heel. The part that is doing all the work to keep the boat in this position is none other than the chainplate. Chainplates are one of the most important parts of the rigging and often the most forgotten about, that is until one of them breaks and the mast becomes unsupported!

External chainplates offer one major advantage over internal chainplates, and that is with the ease of inspection. Having them out and in the open makes them very easy to inspect at any moment. Problems such as cracks or corrosion are readily visible as you look at them every time you look at your topsides.

The two major drawbacks for external chainplates have to do with staying angles and the way that it affects your sheeting angles, and leaks. First, if your chainplates are outside of your hull, then obviously the shrouds will be led to the outside of your hull as well. This means that overlapping headsails will likely need to be sheeted to the toerail behind the shrouds. Outboard sheeting has a direct effect on how far inboard you can sheet the genoa, which directly affects how high you can point into the wind. The angle between the tack and the sheeting position is known as the sheeting angle and the lesser this angle is, the closer to the wind you can sail. If you could sheet the sail on a track on the deck, inboard of the toerail, you would also be able to trim the genoa for sailing closer to the wind. When sailing upwind, especially when tacking upwind, being able to sail closer to the wind will make a huge difference in your VMG, or velocity made good, which is essentially how fast you are sailing to your windward course. If your sailing angle is 30° off the wind, you will get there much faster and sail fewer miles than if you were sailing at 70° off the wind.

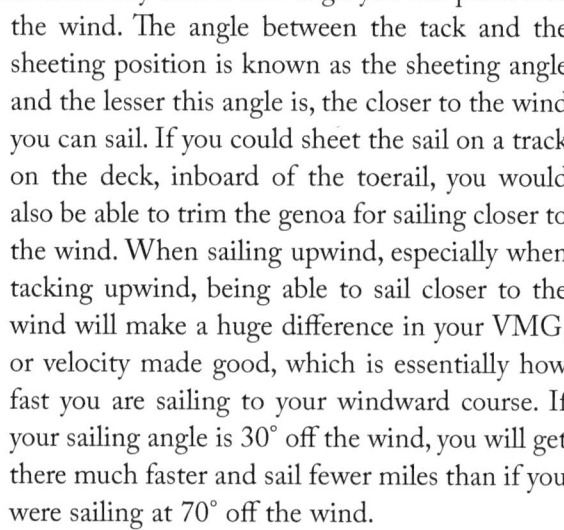

The second major drawback is leaks, as each bolt hole is a potential source of leaks and if you have multiple holes on the topsides that go underwater while sailing, then you now also have multiple potential leak points being submerged. This problem can be largely ignored by both making sure that the bolts are properly bedded with a proper sealant and also by regularly inspecting them for leaks. Attending the issue early rather than letting the problem dribble in for years until it gets "really bad" is a good practice on a sailboat.

Internal chainplates, aside from looking neater, offer some major performance advantages. As mentioned earlier, they don't have to be at the toerail and by moving the shrouds inboard, the sheeting angles can be decreased, allowing you to sail closer to the wind and reach your windward mark even faster.

The biggest structural problem with internal chainplates set inboard of the toerail is increased rig loads. As the shrouds become closer to the mast, they begin to pull more in a vertical direction instead of a lateral direction. This is the purpose of a spreader, to hold the stay out until it's close to the attachment point so that it can approach the mast at an effective angle. By moving the shrouds inboard, everything gets more vertical and the rig loads greatly increase as a result.

The advantage of having fewer holes in the boat is a pseudo improvement. Multiple holes on the side of the hull that do not move is better than one hole on the deck that might wiggle a little bit. Holes on the side of the hull will be exposed to water that is rushing past only when the boat is heeled over. Internal chainplates pop out through the deck and that means that there are holes on the top of the deck that will be exposed to water, and standing water, all the time. If there is spray from waves wetting the deck, the chainplate holes will be exposed. If there is rain, the chainplate holes will be exposed. If there is any breakdown in the sealant of the chainplates, you will have a leak and almost every sailboat has some kind of water damage on the interior from a chainplate that leaked at some point in the boat's history.

Having the chainplates hidden might be a cosmetic improvement to some, but it presents a real struggle when it comes time to inspect them. They tend to be hidden away inside cabinetry or behind the interior woodwork, precluding them from ever being thought of or easily inspected. The most dangerous part of internal chainplates is the section of the chainplate that traverses the hole in the deck, as it is uninspectable without actually unbolting the chainplate and pulling it out to check for cracks or corrosion. This section is also the most susceptible to pitting and crevice corrosion, as the small gap around the failed sealant will become anaerobic and filled with salty water. As the top of the chainplate heats up in the sun, the acidic ocean water around it will also heat up to form a hot acid solution that will accelerate the corrosion process far from view until it is removed for inspection or it finally snaps and breaks free.

Chainplates come in a few different flavors as metallurgy has been evolving over the years. Originally, chainplates were made of iron, but these would rust terribly and by the time yachts came into the picture, chainplates were made of bronze. Bronze is an alloy made by combining copper with other elements. The most common bronze on a boat is silicon bronze, also written as Si bronze. This metal is nonreactive with saltwater, making it a favorite for marine construction. It is a bit soft though and will give you ample warning that the metal structure has been overstressed by bending slightly, indicating that it needs to be replaced. This is much better than holding tightly until it snaps apart leading to a catastrophic failure, which is what other metals have a tendency to do. In the world of chainplates, Si bronze is not the only flavor of bronze, aluminum bronze (Al bronze) and manganese bronze (Mn bronze) are also used. Al bronze is an alloy made of aluminum, iron, and copper and is a very strong alloy thanks to the addition of iron. Mn bronze is incredibly strong thanks to the combination of copper, iron, zinc, aluminum, and manganese.

While Al bronze and Mn bronze are stronger than Si bronze, they have a few caveats that make them the incorrect choice for chainplates. Al bronze will not tarnish like Si bronze because the aluminum reacts with oxygen in the air to form an oxide layer that protects the rest of the metal. This type of bronze relies on the oxide layer for corrosion protection, and if that layer is lost, as will be discussed later in this chapter, the entire piece becomes susceptible to developing crevice corrosion.

Manganese bronze is actually a misnomer because it is technically a brass due to its high zinc content. Initially, Mn bronze is incredibly strong, but when exposed to saltwater, the distance between copper and zinc on the galvanic scale (see p. 167) will cause these two metals to interact. The result is called dezincification, a process where the zinc in the alloy strips away, leaving the metal as a brittle sponge of what it once was. This is why Mn bronze is incredibly strong when first fabricated but it will become brittle as time progresses. For these reasons, the alloy of choice that will last several lifetimes is none other than Si bronze.

As wonderful as bronze is on a boat, it has one major drawback: It tarnishes. Si bronze will develop a nice green patina on its surface rather quickly. For a work boat or a salty cruiser, this green layer is a badge of honor; but for the yachty who wants everything to be bright and shiny, this green layer is a constant source of labor. Polishing bronze to keep it bright has led to the term "brightwork," which now has been expanded upon to describe varnished or oiled wood on the boat's deck and hull as well. Brightwork is work and the only way to keep something bright without the work associated is to choose a different metal. Due to the trend in aesthetics, bronze quickly faded away to be replaced by stainless steel, which holds polish very well and for a long time, greatly reducing the amount of labor involved in maintaining the brightwork on a boat.

The three most common stainless-steel alloys you will find on a boat are 304, 316, and 316L. 304 stainless steel contains a minimum of 66% iron, 17% chromium, and 8% nickel, and then smaller amounts of other elements. 316 stainless steel is similar in composition to 304, with the addition of molybdenum. The difference between 316 and 316L is that 316 contains no more than 0.08% carbon, while 316L contains no more than 0.03% carbon.

304 stainless steel has superior rust protection over regular carbon steel, making it "stain-less" for household applications. The problem is a sailboat is anything but

"household application." Cheap parts from household hardware stores labeled "stainless" are most likely going to be made out of 304, and this fact will become readily apparent when they burst into rust after a short period of time.

The addition of molybdenum makes the alloy much more resistant to saltwater corrosion, which is why 316 is often referred to as "marine grade stainless steel." While 316 is more rust resistant, it is also harder to work with, which makes fabricating parts with this alloy even more expensive. 316 is the alloy of choice for all fasteners above the waterline on a sailboat as they are incredibly strong and will not rust for many years. Special care does need to be taken when using stainless steel fasteners into an aluminum spar as the dissimilar metals will cause galvanic corrosion to eat away at the least noble metal in a hurry, quickly turning the least noble metal into an oxide dust of the metal it once was.

Reducing the carbon content of 316 yields 316L, which has very similar strength properties but greatly improved rust protection over regular 316. The leap in rust protection between steel to 304 stainless steel is dramatic, and the next leap from 304 to 316 is equally impressive. Going one step further into the realm of rust avoidance with 316L is a wonderful thing on a sailboat that will be forced to endure its lifetime in the marine environment.

The way stainless steel works to prevent rusting is a little bit magical. Carbon steel is pretty much iron with a smidge of carbon added in there to transform it into a metal that has revolutionized humanity. Iron quickly reacts with oxygen to form iron oxide, or "rust." The only way to prevent iron from rusting is to keep oxygen away from it. This can be accomplished by using some form of coating to keep the iron isolated from the oxygen in the air, thereby inhibiting the chemical reaction that forms iron oxide. Traditional methods involve coating the metals surface with oil or paint, but stainless steel takes it to a whole new level. The high levels of chromium in the stainless steel actually react with oxygen before the iron can, and forms an extremely thin layer of chromium oxide on the surface. This thin layer protects the iron inside and prevents the whole part from rusting.

316 and 316L with their low corrosion risk and incredible strength are only made even better by the fact that they can be polished to a mirror finish, which allows the brightwork on your boat to remain bright and maintenance free! As long as the metal is freely exposed to a constant supply of oxygen, the oxide layer will remain intact on the surface of the metal and no corrosion will occur, even as waves come crashing onto the deck, dousing everything with a nice helping of corrosive seawater!

Using 316L for chainplates seems like the most logical choice for metals, which is why pretty much every single sailboat you will see has this specific alloy employed for their chainplates to support the rig and transfer the load to the hull beneath. Stainless steel sounds like someone with a face so beautiful you could stare into their eyes forever—until you notice the mole on their left cheek. This isn't just any mole, because this one has thick curly hairs growing out of it!

The only blemish to the stainless persona is that if the alloy doesn't have a steady flow of fresh oxygen swirling around it, that protective oxide will disappear and leave the chainplate as defenseless as a starship with its shields down in every sci-fi movie. With the loss of the oxide layer, the stainless steel becomes susceptible to crevice corrosion, which is a very difficult type of corrosion to find and can grow unnoticed for a long time.

Crevice corrosion starts off as small pits in the surface that occur as soon as the protective oxide layer is lost. Chloride ions, which are present in seawater, accumulate in the pitted surface and then acidify the environment, which promotes the breakdown of the metal through anodic dissolution. As time progresses, the small pits turn into small crevices that can run through the metal with no outward signs other than a microscopic crack somewhere on the surface.

As you might be thinking, if you can stop the pitting, you can then prevent the crevice corrosion that follows. It turns out that the elemental components of the stainless steel have the biggest effect on preventing pitting in the first place. Chromium and molybdenum are the two best elements to prevent pitting, which is why having one is good and having both is better, marking the difference between 304 and 316.

The first step in preventing pitting is to maintain a constant flow of oxygen around the metal. The reality is, this is very difficult to do since the chainplate needs to be bolted to something and more especially, if the chainplate is internal, needs to pass through the deck where it is bedded down with sealant to keep water from entering. When the sealant breaks down and small gaps begin to form where seawater can pool in this oxygen-poor environment, the chainplate will begin to pit and the entire process begins.

You might now be wondering, "Why on earth would they use stainless steel for chainplates if it suffers from crevice corrosion?" Everything on a boat has a finite lifespan and the stainless steel of the chainplate lasts about ten years. After this period of time, chances are one of your chainplates has developed crevice corrosion and is at risk of breaking, taking your rig with it. This is why insurance companies require you to replace your standing rigging every ten years to stay insured. They have done their calculations and have found that the risk of the rig failing in under ten years is rare, but the risk of you submitting a claim increases exponentially after the ten-year mark.

Stainless steel chainplates are strong, shiny, and going to last a known amount of time. This predictable outcome helps make them a reliable choice for chainplate materials. While 316L is the standard alloy for chainplates, it is not the only metal that is being employed for such uses.

Titanium has made quite a splash as it entered the sailing scene. Titanium is stronger and lighter than stainless steel, meaning that stainless steel parts can be replaced with a smaller and even lighter product that will get the same job done. In the world of high-performance sailing, where every bit of weight counts, the prospect of shaving a significant amount of weight by using a different metal is music to a foiling yacht's ears. Titanium also exhibits better corrosion resistance than stainless steel, making it seem like the best metal to be found on a sailboat.

While titanium does seem like the wonder metal to save us from the pitfalls of stainless steel, you need to make sure it is the right grade of titanium for the purpose. Titanium develops its incredible corrosion resistance thanks to the oxide layer that forms on the surface of the metal. Just like with stainless steel, if it doesn't have enough fresh oxygen floating around, that oxide layer will degrade away and the metal will be unprotected; but unlike stainless steel, the oxide layer doesn't break down until certain conditions are met. Pitting and crevice corrosion are possible in unalloyed titanium when exposed to seawater and temperatures above 180°F (82°C). Grades 7 and 12 offer resistance to pitting and crevice corrosion when exposed to seawater at temperatures as high as 500°F (260°C). While 180°F (82°C) sounds hotter than any condition that the boat will experience, just think about how hot the gray chainplate will get when baking in the tropical sun. Add a little splashing from your last sail to make a bit of water pool around the metal, and you now have a very hot environment with the presence of seawater. Grades 7 and 12 use other metals in the alloy to increase the corrosion resistance of the metal, but if the wrong grade of titanium is used, it will suffer an early death just like if the wrong grade of steel were selected. Just because it is "titanium" doesn't mean it is the best; it has to be the right alloy or it will work as well as any other randomly selected metal.

While titanium really seems like the ultimate material for metal chainplates, they have one major drawback that does not look like it will be overcome anytime soon: price. Titanium chainplates are incredibly expensive! The price of titanium to fabricate the chainplate is about five times the price of the equivalent stainless steel. This means that the cost for chainplates will be about five times as high!

So far, every option for a chainplate has involved some kind of metal that needs to be fitted to the boat to transmit the forces appropriately, and the metal seems to be the weak point in every system presented. What about removing the metal and having the rigging connect directly to the hull? This is the concept behind composite chainplates, where the hull layup includes directional fibers that are purposefully designed to spread the load out over the desired area. There is no metal to corrode, tarnish, rust, crack, or any other issue that plagues today's options for chainplate materials. By incorporating the chainplate into the hull layup design, the forces of the rigging can be spread to exactly the right areas and in the right amounts. Best of all, the chainplates are actually the hull itself so there is no concern of the chainplate breaking off and coming out through the deck! The other advantage of this is it offers even greater weight savings. The lightest of metals still need to be fastened to a reinforced area of the hull, which then has to spread the load accordingly. By removing the physical chainplate, the hull itself simply forms a method to connect to the rigging, usually by incorporating a tang, where the turnbuckle or deadeye can attach.

While composite chainplates sound like the best option of all, they are not one that is easily retrofitted. The chainplate laminate needs to be incorporated into the hull and if you are retrofitting, then you either need to strip out the interior to expose the hull to begin the layup of this large structure or you will need to make it external, which will then necessitate a new topside paint job to finish the hull. Both of these options are costly and involved. Switching a metal chainplate from one alloy to a better one is a much easier method to improve the longevity and characteristics of your chainplates without having to undergo major modification to your yacht.

If you are able to access the space inside your boat, retrofitted composite chainplates need to be custom made to be an exact fit to the inside of your hull. The carbon fiber is then sized appropriately to be much stronger than necessary simply because it is easy to overbuild and then not worry about it for the rest of the boat's life. In general, metal chainplates are four-thirds as strong as the wire rigging they support, giving you a one-third margin of safety. For carbon fiber chainplates, the goal is to make them 20x as strong as the wire they support! To accomplish this level of overkill in metal would incur both a cost and weight penalty, as the chainplates would be unruly to manage during installation, not to mention the weight penalty for the life of the yacht. Carbon fiber chainplates weigh barely anything so overbuilding them doesn't carry along the same penalties.

The greatest advantage of composite chainplates is that they become an integral part of the hull. The forces are transmitted directly onto structures in a planned and calculated manner without the use of fasteners that can fail or metals that can degrade over time. Creep in the material is also ignored as the fibers of the composite chainplate will never really be strained anywhere near their limits during their life. The concern of any leaks also fades away as the chainplates simply emerge as part of the deck as a small ridge that the stay attaches to via a pinned joint. As time marches on, you will see more and more boats being designed with these types of chainplates until they simply become commonplace and the archaic methods of bolting a piece of metal to the hull is read about in historical books.

To summarize, the most common chainplate material is 316L, while the ideal chainplate material is either Si bronze, grade 7 titanium, or grade 12 titanium (other grades of titanium are inferior and won't offer the same advantage that you think you will be getting). Composite chainplates negate all the problems of metal chainplates by incorporating their fibers and structure into the hull layup, thus providing the ultimate setup for chainplates.

The final part of the chainplate story is going to be sizing the chainplate accordingly. In general, chainplates should be sized to have a tensile or yield strength, depending on the material, that is at least four-thirds the strength of the rigging. If the rig gets overloaded and something is going to break, you don't want the "fuse" in the system to be the chainplate because that is not an easy fix. Ideally, the clevis pin will break first, shearing under the intense load and letting the rig fall to protect all the other components of the boat. Since the clevis pin is determining the maximum load that the rig will support, there is really no point in overbuilding the chainplates. Remember, the bigger they get, the more expensive they become, and cash is a finite resource that needs to make its way around the entire boat to keep it floating and looking beautiful.

To figure out what size your chainplate needs to be, you can reference the table below, which uses the standing rigging wire size to help set the stage for the appropriate chainplate size. These are the safe minimum sizes that you can use, but you can always make the chainplate bigger, within reason, and therefore more robust. For metal chainplates, the length is dependent on how many fasteners you need to employ to firmly and safely attach the chainplate to the boat. For composite chainplates, which are bonded to the hull, their size is dependent upon the bond strength of the epoxy that will attach them to the boat. The strength rating for carbon fiber and epoxy is listed to ensure that the material you are using is at least as strong as the one that is calculated in the table.

The breaking strength of your chainplate will then determine the number of bolts you will need to use. Using a larger bolt will let you use fewer bolts, and going smaller will necessitate more bolts. The strength of each bolt can be found in the table on page 111. To figure out how many bolts you need, divide the breaking strength of the wire you are using by the strength of your selected bolt size and round up to the nearest whole number. The length of the bolt is also important as the threaded part should not be in contact with the chainplate, that way the most metal is employed. These calculations have a few safety factors built in, first the cross-sectional area is the threaded part, so that will undervalue the strength of the bolt. Second, it assumes that the bolt will be in shear load, which is not the actual case. If you don't tighten your bolt enough, it will be loaded in shear, but when properly torqued, it will be under tension, which is twice as strong as the shear rating used here. The goal of the torque spec is to create enough clamping pressure that the bolt is acting as a clamp and not actually supporting the load in shear. The chainplate is actually held by the friction of the metal onto the mating surface in what is known as a slip-critical joint. Each bolt material, size, and thread count has its own particular torque spec, which can be provided by the manufacturer. For 18-8 stainless and bronze, the table below can help.

Stainless Steel Wire Diameter	651 Si Bronze									
	Breaking Strength		Clevis Pin Size	Chainplate Thickness		Chainplate Width		Clevis Pin Hole Center below Top	
	lbs	kg	inches	inches	mm	inches	mm	inches	mm
3/32	1067	485	1/4	1/8	3.5	3/4	19	0.59	16
3mm	1666	757	1/4	1/8	3.5	3/4	19	0.59	16
1/8	1898	863	1/4	1/8	3.5	3/4	19	0.59	16
9/64	2348	1067	1/4	1/8	3.5	15/16	24	0.59	16
5/32	2965	1348	5/16	3/16	5	15/16	24	0.74	19
4mm	2965	1348	5/16	3/16	5	15/16	24	0.74	19
3/16	4268	1940	3/8	3/16	5	1-1/8	29	0.89	23
5mm	4632	2105	3/8	1/4	6.5	1-1/8	29	0.89	23
7/32	5811	2641	7/16	1/4	6.5	1-5/16	34	1.04	27
6mm	6671	3032	1/2	1/4	6.5	1-1/2	39	1.19	31
1/4	7588	3449	1/2	1/4	6.5	1-1/2	39	1.19	31
7mm	9076	4125	1/2	5/16	8	1-1/2	39	1.19	31
9/32	9605	4366	1/2	5/16	8	1-1/2	39	1.19	31
5/16	11858	5390	5/8	5/16	8	1-7/8	48	1.48	38
8mm	11858	5390	5/8	5/16	8	1-7/8	48	1.48	38
9mm	15009	6822	5/8	3/8	10	1-7/8	48	1.48	38
3/8	17076	7762	5/8	7/16	12	1-7/8	48	1.48	38
10mm	18528	8422	3/4	1/2	13	2-1/4	58	1.78	46
11mm	22419	10190	3/4	1/2	13	2-1/4	58	1.78	46
7/16	23241	10564	3/4	1/2	13	2-1/4	58	1.78	46
12mm	26680	12127	7/8	1/2	13	2-5/8	67	2.08	53
1/2	29883	13583	7/8	1/2	13	3	77	2.08	53
13mm	31312	14233	7/8	1/2	13	3	77	2.08	53
14mm	36316	16507	15/16	5/8	16	3-3/4	96	2.23	57
9/16	37888	17222	15/16	5/8	16	3-3/4	96	2.23	57
15mm	41689	18950	1	5/8	16	4-1/2	115	2.38	61
5/8	45194	20543	1-1/16	5/8	16	4-7/8	124	2.52	65
3/4	51717	23508	1-1/8	5/8	16	5-1/4	134	2.67	68

316 Stainless Steel									
Stainless Steel Wire Diameter	Breaking Strength		Clevis Pin Size	Thickness		Width		Clevis Pin Hole Center below Top	
	lbs	kg	inches	inches	mm	inches	mm	inches	mm
3/32	1067	485	1/4	1/8	3.5	3/4	19	0.59	16
3mm	1666	757	1/4	1/8	3.5	3/4	19	0.59	16
1/8	1898	863	1/4	1/8	3.5	3/4	19	0.59	16
9/64	2348	1067	1/4	1/8	3.5	15/16	24	0.59	16
5/32	2965	1348	5/16	3/16	5	15/16	24	0.74	19
4mm	2965	1348	5/16	3/16	5	15/16	24	0.74	19
3/16	4268	1940	3/8	3/16	5	1-1/8	29	0.89	23
5mm	4632	2105	3/8	1/4	6.5	1-1/8	29	0.89	23
7/32	5811	2641	7/16	1/4	6.5	1-5/16	34	1.04	27
6mm	6671	3032	1/2	1/4	6.5	1-1/2	39	1.19	31
1/4	7588	3449	1/2	1/4	6.5	1-1/2	39	1.19	31
7mm	9076	4125	1/2	5/16	8	1-1/2	39	1.19	31
9/32	9605	4366	1/2	5/16	8	1-1/2	39	1.19	31
5/16	11858	5390	5/8	5/16	8	1-7/8	48	1.48	38
8mm	11858	5390	5/8	5/16	8	1-7/8	48	1.48	38
9mm	15009	6822	5/8	3/8	10	1-7/8	48	1.48	38
3/8	17076	7762	5/8	7/16	12	1-7/8	48	1.48	38
10mm	18528	8422	3/4	1/2	13	2-1/4	58	1.78	46
11mm	22419	10190	3/4	1/2	13	2-1/4	58	1.78	46
7/16	23241	10564	3/4	1/2	13	2-1/4	58	1.78	46
12mm	26680	12127	7/8	1/2	13	2-5/8	67	2.08	53
1/2	29883	13583	7/8	1/2	13	3	77	2.08	53
13mm	31312	14233	7/8	1/2	13	3	77	2.08	53
14mm	36316	16507	15/16	5/8	16	3-3/4	96	2.23	57
9/16	37888	17222	15/16	5/8	16	3-3/4	96	2.23	57
15mm	41689	18950	1	5/8	16	4-1/2	115	2.38	61
5/8	45194	20543	1-1/16	5/8	16	4-7/8	124	2.52	65
3/4	51717	23508	1-1/8	5/8	16	5-1/4	134	2.67	68

316L Stainless Steel									
Stainless Steel Wire Diameter	Breaking Strength		Clevis Pin Size	Thickness		Width		Clevis Pin Hole Center below Top	
	lbs	kg	inches	inches	mm	inches	mm	inches	mm
3/32	1067	485	1/4	1/8	3.5	1	26	0.59	16
3mm	1666	757	1/4	1/8	3.5	1	26	0.59	16
1/8	1898	863	1/4	1/8	3.5	1	26	0.59	16
9/64	2348	1067	1/4	1/8	3.5	1-1/4	32	0.59	16
5/32	2965	1348	5/16	3/16	5	1-1/4	32	0.74	19
4mm	2965	1348	5/16	3/16	5	1-1/4	32	0.74	19
3/16	4268	1940	3/8	3/16	5	1-3/4	45	0.89	23
5mm	4632	2105	3/8	1/4	6.5	1-3/4	45	0.89	23
7/32	5811	2641	7/16	1/4	6.5	1-3/4	45	1.04	27
6mm	6671	3032	1/2	5/16	8	2	51	1.19	31
1/4	7588	3449	1/2	3/8	10	2	51	1.19	31
7mm	9076	4125	1/2	3/8	10	2	51	1.19	31
9/32	9605	4366	1/2	3/8	10	2	51	1.19	31
5/16	11858	5390	5/8	1/2	13	2	51	1.48	38
8mm	11858	5390	5/8	1/2	13	2	51	1.48	38
9mm	15009	6822	5/8	1/2	13	2-1/2	64	1.48	38
3/8	17076	7762	5/8	1/2	13	2-3/4	70	1.48	38
10mm	18528	8422	3/4	5/8	16	3	77	1.78	46
11mm	22419	10190	3/4	5/8	16	3	77	1.78	46
7/16	23241	10564	3/4	5/8	16	3	77	1.78	46
12mm	26680	12127	7/8	3/4	20	3-1/4	83	2.08	53
1/2	29883	13583	7/8	3/4	20	3-1/4	83	2.08	53
13mm	31312	14233	7/8	3/4	20	3-1/2	89	2.08	53
14mm	36316	16507	15/16	3/4	20	3-3/4	96	2.23	57
9/16	37888	17222	15/16	3/4	20	3-3/4	96	2.23	57
15mm	41689	18950	1	7/8	23	3-3/4	96	2.38	61
5/8	45194	20543	1-1/16	15/16	24	4	102	2.52	65
3/4	51717	23508	1-1/8	1	26	4-1/2	115	2.67	68

Stainless Steel Wire Diameter	Grade 7 and Grade 12 Titanium								
	Breaking Strength		Clevis Pin Size	Thickness		Width		Clevis Pin Hole Center below Top	
	lbs	kg	inches	inches	mm	inches	mm	inches	mm
3/32	1067	485	1/4	1/8	3.5	3/4	20	0.59	16
3mm	1666	757	1/4	1/8	3.5	3/4	20	0.59	16
1/8	1898	863	1/4	1/8	3.5	3/4	20	0.59	16
9/64	2348	1067	1/4	1/8	3.5	3/4	20	0.59	16
5/32	2965	1348	5/16	1/8	3.5	15/16	24	0.74	19
4mm	2965	1348	5/16	1/8	3.5	15/16	24	0.74	19
3/16	4268	1940	3/8	3/16	5	1-1/8	29	0.89	23
5mm	4632	2105	3/8	3/16	5	1-1/8	29	0.89	23
7/32	5811	2641	7/16	3/16	5	1-5/16	34	1.04	27
6mm	6671	3032	1/2	3/16	5	1-1/2	38.5	1.19	31
1/4	7588	3449	1/2	3/16	5	1-1/2	38.5	1.19	31
7mm	9076	4125	1/2	1/4	6.5	1-1/2	38.5	1.19	31
9/32	9605	4366	1/2	1/4	6.5	1-1/2	38.5	1.19	31
5/16	11858	5390	5/8	1/4	6.5	1-7/8	48	1.48	38
8mm	11858	5390	5/8	1/4	6.5	1-7/8	48	1.48	38
9mm	15009	6822	5/8	5/16	8	1-7/8	48	1.48	38
3/8	17076	7762	5/8	3/8	10	1-7/8	48	1.48	38
10mm	18528	8422	3/4	3/8	10	2-1/4	57.5	1.78	46
11mm	22419	10190	3/4	3/8	10	2-1/4	57.5	1.78	46
7/16	23241	10564	3/4	3/8	10	2-1/4	57.5	1.78	46
12mm	26680	12127	7/8	7/16	11.5	2-5/8	67	2.08	53
1/2	29883	13583	7/8	7/16	11.5	2-5/8	67	2.08	53
13mm	31312	14233	7/8	7/16	11.5	2-5/8	67	2.08	53
14mm	36316	16507	15/16	1/2	13	2-13/16	71.5	2.23	57
9/16	37888	17222	15/16	1/2	13	2-13/16	71.5	2.23	57
15mm	41689	18950	1	1/2	13	3	76.5	2.38	61
5/8	45194	20543	1-1/16	9/16	14.5	3-3/16	81	2.52	65
3/4	51717	23508	1-1/8	9/16	14.5	3-3/8	86	2.67	68

Stainless Steel Wire Diameter	Breaking Strength		Clevis Pin Size	Thickness		Width		Clevis Pin Hole Center below Top		Bonding Surface Area			
										length		width	
	lbs	kg	inches	inches	mm	inches	mm	inches	mm	inches	mm	inches	mm
3/32	1067	485	1/4	1/16	1.6	3	76.2	0.06	1.59	3	76.2	3	76.2
3mm	1666	757	1/4	1/16	1.6	3	76.2	0.06	1.59	3	76.2	3	76.2
1/8	1898	863	1/4	1/16	1.6	3	76.2	0.06	1.59	3	76.2	3	76.2
9/64	2348	1067	1/4	1/16	1.6	3	76.2	0.06	1.59	4	101.6	3	76.2
5/32	2965	1348	5/16	1/16	1.6	3	76.2	0.06	1.59	5	127.0	3	76.2
4mm	2965	1348	5/16	1/16	1.6	3	76.2	0.06	1.59	6	152.4	3	76.2
3/16	4268	1940	3/8	1/8	3.2	3	76.2	0.13	3.18	6	152.4	3	76.2
5mm	4632	2105	3/8	1/8	3.2	3	76.2	0.13	3.18	7	177.8	3	76.2
7/32	5811	2641	7/16	1/8	3.2	3	76.2	0.13	3.18	9	228.6	3	76.2
6mm	6671	3032	1/2	1/8	3.2	3	76.2	0.13	3.18	10	254.0	3	76.2
1/4	7588	3449	1/2	1/8	3.2	3	76.2	0.13	3.18	11	279.4	3	76.2
7mm	9076	4125	1/2	3/16	4.8	4	101.6	0.19	4.76	12	304.8	4	101.6
9/32	9605	4366	1/2	3/16	4.8	4	101.6	0.19	4.76	11	279.4	4	101.6
5/16	11858	5390	5/8	3/16	4.8	5	127.0	0.19	4.76	11	279.4	5	127.0
8mm	11858	5390	5/8	3/16	4.8	5	127.0	0.19	4.76	11	279.4	5	127.0
9mm	15009	6822	5/8	1/4	6.4	5	127.0	0.25	6.35	13	330.2	5	127.0
3/8	17076	7762	5/8	5/16	7.9	5	127.0	0.31	7.94	15	381.0	5	127.0
10mm	18528	8422	3/4	5/16	7.9	5	127.0	0.31	7.94	16	406.4	5	127.0
11mm	22419	10190	3/4	3/8	9.5	5	127.0	0.38	9.53	19	482.6	5	127.0
7/16	23241	10564	3/4	3/8	9.5	5	127.0	0.38	9.53	20	508.0	5	127.0
12mm	26680	12127	7/8	7/16	11.1	6	152.4	0.44	11.11	19	482.6	6	152.4
1/2	29883	13583	7/8	1/2	12.7	6	152.4	0.50	12.70	21	533.4	6	152.4
13mm	31312	14233	7/8	1/2	12.7	6	152.4	0.50	12.70	22	558.8	6	152.4
14mm	36316	16507	15/16	1/2	12.7	7	177.8	0.50	12.70	23	584.2	7	177.8
9/16	37888	17222	15/16	1/2	12.7	7	177.8	0.50	12.70	23	584.2	7	177.8
15mm	41689	18950	1	1/2	12.7	8	203.2	0.50	12.70	23	584.2	8	203.2
5/8	45194	20543	1-1/16	1/2	12.7	8	203.2	0.50	12.70	24	609.6	8	203.2
3/4	51717	23508	1-1/8	1/2	12.7	10	254.0	0.50	12.70	24	609.6	10	254.0

Carbon Fiber
Tensile Strength: 600,000psi or 4,136MPa
Epoxy Bond Strength: 6,400psi or 0.027MPa

DOWN TO THE BOTTOM

Bolt Size	Chainplate Bolts							
	Cross Sectional Area		Si Bronze Shear Strength 38,000psi or 262MPa		316 Shear Strength 45,000psi or 310MPa		18-8 Shear Strength 30,000psi or 205MPa	
	in²	mm²	lbs	kg	lbs	kg	lbs	kg
M4-.7	0.0082	5.3479	311.60	141.64	369.00	167.73	246.00	111.82
M6-1	0.0307	19.8444	1,166.60	530.27	1,381.50	627.95	921.00	418.64
1/4-20	0.0318	20.516	1,208.40	549.27	1,431.00	650.45	954.00	433.64
1/4-28	0.0364	23.4838	1,383.20	628.73	1,638.00	744.55	1,092.00	496.36
5/16-18	0.0525	33.8709	1,995.00	906.82	2,362.50	1,073.86	1,575.00	715.91
5/16-24	0.0581	37.4837	2,207.80	1,003.55	2,614.50	1,188.41	1,743.00	792.27
M8-1.25	0.0634	40.9561	2,409.20	1,095.09	2,853.00	1,296.82	1,902.00	864.55
3/8-16	0.0775	48.7095	2,945.00	1,338.64	3,487.50	1,585.23	2,325.00	1,056.82
3/8-24	0.0878	56.645	3,336.40	1,516.55	3,951.00	1,795.91	2,634.00	1,197.27
7/16-14	0.1063	68.5805	4,039.40	1,836.09	4,783.50	2,174.32	3,189.00	1,449.55
M10-1.5	0.1064	68.6771	4,043.20	1,837.82	4,788.00	2,176.36	3,192.00	1,450.91
1/2-13	0.1419	91.5482	5,392.20	2,451.00	6,385.50	2,902.50	4,257.00	1,935.00
M12-1.75	0.1594	102.8557	6,057.20	2,753.27	7,173.00	3,260.45	4,782.00	2,173.64
5/8-11	0.2261	145.8706	8,591.80	3,905.36	10,174.50	4,624.77	6,783.00	3,083.18
3/4-10	0.3346	215.8705	12,714.80	5,779.45	15,057.00	6,844.09	10,038.00	4,562.73

CHAPTER NINE
Rigging Materials

THE SHAPE AND SIZE OF THE RIGGING IS LIMITED ONLY BY OUR IMAGINATION AND the materials that we have presently to rig with. Imagination has driven the development of innovative materials that have revolutionized the way rigging is set up on a sailboat, but we need to physically have these materials present in order to make it the norm. I'm sure that Nordic shipwrights wanted to build taller masts to support larger sails, but the material limitations of the time did not allow for such towering designs to be able to sail across the ocean safely.

For thousands of years, rigging used natural materials to achieve the desired effects. Sails were made out of cotton and other such fibers woven into textiles, while running and standing rigging was made out of hemp, manila, or other such plants that had fibrous parts that could be employed for man's uses. These spun fibers could be turned into ropes that ranged in duty from supporting the mast to hanging from the bell's clapper. Natural fibers are subject to rot if they get wet for too long, and being on a sailboat, the time of its existence was passing like grains of sand through an hourglass. Since the same material was used in various parts of the boat, large stores of the rope could be kept dry and at the ready to replace decaying sections on long passages.

The rope made from this type of fiber is called three-lay or three-strand as it consists of three individual pieces of spun cord that are then twisted together to form the rope. This type of rope can be repaired easily, with a variety of mending splices where new sections of spun cord can be laid into the line to replace a part that has been damaged. The serviceability of this type of rope does make you wonder if that is why this type of rope was used on vessels for so many centuries. Any port you sailed to, this type of rope would be available and repair or replacement of any part of the rigging was possible.

As materials improved beyond these naturally occurring fibers, the specialization of the different parts of rigging began to emerge. Standing rigging switched from rope to wire, and galvanized steel rigging was a popular rigging material for the first half of the twentieth century. Steel rusts very quickly on a sailboat, but if the metal could be covered in a layer of zinc, rust could be prevented and the steel beneath could last a long time. Galvanized wire was twisted into a rope, also referred to as "wire rope," which could be spliced and used in much the same way as the hemp rope rigging that predated it. The steel was protected for as long as the galvanization over it remained intact. For this purpose, various methods of preserving the wire were devised but the most successful was to "worm, parcel, and serve" the wire and then paint it with "slurry."

Worm, parcel, and serve refers to a method of covering the wire completely, thus isolating it from the corrosive marine environment. Let's break down each step of this system.

The worm part refers to laying a very small line into the valleys between the galvanized rope strand lays. This converts the wire rope from having high and low spots on its surface to being more of an even circle.

Parceling refers to laying and wrapping tarred canvas over the wormed wire rope. The tar helps seal up the galvanized wire and protect it from the elements.

Servicing is when you tightly wrap a very thin line around the outside of the entire setup to seal it up nice and tight. The thin line actually becomes much stronger when wrapped tightly, imparting even more protection to the galvanized wire beneath. If the worm step is not performed, the serving would have high points caused by the wire rope, which will cause chafe to start there. By laying the worming in the grooves of the lay, the entire surface is much smoother and rounder, making the whole structure more resistant to chafe.

Over time, the tar in the parcel layer will dry out, which is why it is imperative that the juiciness of it all be replenished. The drink of choice is called slurry, and it rejuvenates the parcel layer to its original glory. The exact recipe for slurry is a bit nebulous as each salty sailor had their own "perfect recipe," but the general ingredients are pine tar, turpentine, varnish, cooked bacon grease, pork chop drippings, and paint. The general principle is to soak the stay in some greasy goodness that will keep water out and oil in. If well maintained, galvanized rigging could live indefinitely in this cocoon of oily paste. The problem is that this is all being done high above the boat where all the little drips will make their way onto the deck. Black goo falling from the sky will stain those white pants everyone imagines a yachtie wears.

By the 1960s, stainless steel rigging was starting to catch on with sailing yachts and the sun was setting on the days of galvanized rigging. Stainless steel rigging is available in both 304 and 316, and as previously discussed, they have different levels of resistance to corrosion. Stainless steel wire does not last nearly as long as galvanized wire that is wormed, parceled, and served, and maintained with slurry, but it is clean,

RIGGING MATERIALS

shiny, and maintenance free by comparison. New terminators became commonplace on sailboats as stainless steel wire was seized into a fitting instead of spliced around an eye. Splicing wire rope is a very laborious process that will definitely result in you getting stabbed by uncooperative wires at some point, whereby pinching the wire into the end of a terminator could be accomplished quickly and painlessly. The result is also much sleeker looking as the wire simply ends with a nice-looking stainless steel fitting that leads into a turnbuckle.

The evolution of materials has come a long way, but in the grand scheme of things, it was still rope. Hemp was made out of three strands wrapped around each other. Galvanized wire is called 7×7 because it is made of seven smaller strands that are twisted to make a strand, and then seven of those strands are twisted to make the rope. Stainless steel wire uses a pattern referred to as 1×19 where each strand is made out of one single wire, and those 19 wires are wrapped around each other.

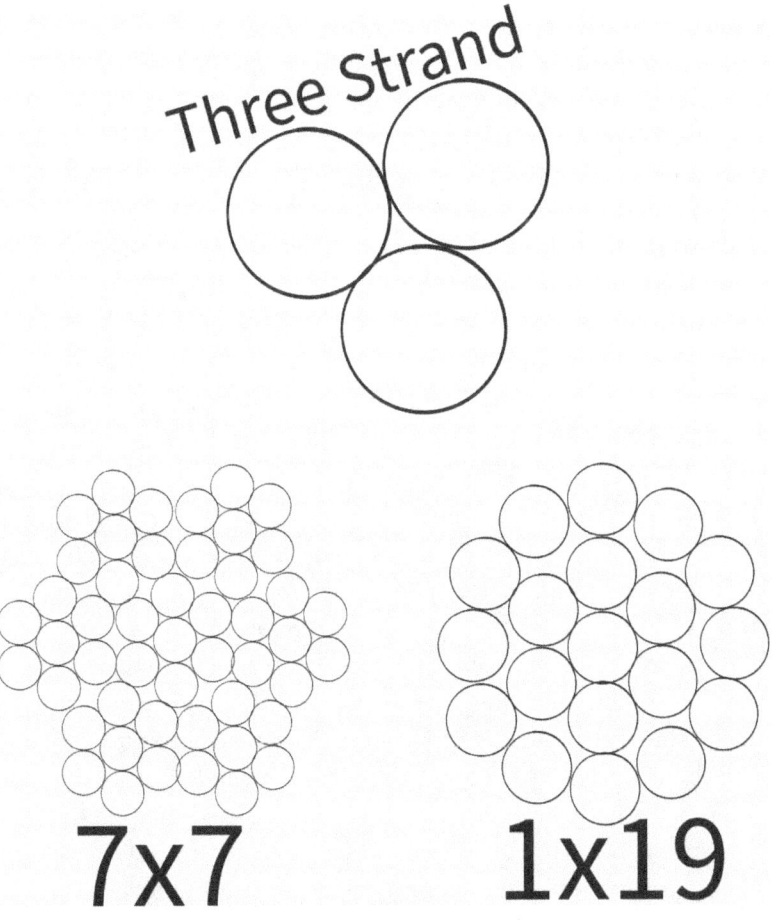

The first real departure from history came in the 1970s with rod rigging. For the first time, the mast was not supported by some type of rope, but instead by a single stainless steel rod that runs from turnbuckle to mast fitting. There are no twists or turns, as it is one single, straight piece of metal. This type of rigging was revolutionary and quickly gained favor on racing sailboats as the thinner diameter of the stay meant that it was both less windage and less weight aloft. The performance advantages along with the simplicity and clean appearance of rod rigging brought it to the forefront of sailboat rigging.

The reason that 1×19 wire rigging didn't disappear the same way that 7×7 galvanized wire did is because rod rigging has a major design flaw which most sailors are not willing to risk. To secure the rod into the terminators on the mast or turnbuckle, the rod needs to be shaped into a head. The process of shaping the rod causes the metal to work harden and develop stress cracks. These cracks are barely visible and occur inside of the terminator, in other words, you can't see if the head is cracking off unless you disassemble the rigging and X-ray it to evaluate for signs of cracks. Unlike stainless steel that has a working life of around ten years, rod rigging can start cracking the moment it is made. This means that brand new or really old rod rigging is always at risk of breaking its head off and releasing the mast from its grip. Most sailors are not willing to risk this level of catastrophic failure as they want rigging that they can depend on and not rigging that will give them a slight edge over their competition but might also cost them their rig.

For most of the history of sailing, rigging has been some form or rope. Steel came into the picture for a brief moment, but as technology continues to improve, it seems that rope has returned onto the sailing scene. Synthetic or composite rigging has created a major upset in the dominance of steel thanks to the major weight savings aloft. The current types of synthetic rigging are PBO, carbon fiber, Aramid, and HMPE.

PBO is short for polybenzoxazole and is also called the much easier to remember name Zylon. This fiber will not creep and is the ultimate in synthetic rigging fibers as far as strength and performance goes. As perfect as this fiber may seem, it has two major flaws: First, it degrades very quickly when exposed to UV light meaning that it must be maintained in an intact opaque covering in order to survive. The second problem is that PBO is incredibly expensive. If the cover becomes compromised, then the whole stay needs to be replaced for a substantial sum of money.

Carbon fiber rigging uses long parallel chains of carbon atoms that are formed into filaments. This type of fiber is far stronger than its steel counterpart, allowing the use of much thinner stays, which greatly reduces the windage caused by the rigging.

Aramid is a synthetic fiber made by DuPont and used in their Kevlar. This fiber is incredibly strong, but like PBO needs to be protected from UV exposure. This is another fiber that is typically found on high-performance yachts.

RIGGING MATERIALS

HMPE or UHMWPE is much more affordable than all the other fibers, which has created a huge upset in the world of steel rigging for average sailors and cruisers. HMPE stands for High Modulus Polyethylene and UHMWPE stands for Ultra-High Molecular Weight Polyethylene. These are two different names for the same thing, as they all refer to Dyneema: a fancy plastic that has more strength than steel and a price that is comparable enough to gain the attention of everyday sailors who want something better. Dyneema comes in several classes, each exhibiting its own characteristics that will be discussed further. Considering how PBO, Aramid, and carbon fiber rigging are all cost-prohibitive, we will focus the rest of our time on Dyneema, which is the type of material you are most likely to encounter if you meet a sailor who has synthetic rigging on their boat.

The thing that all synthetic rigging systems have in common is that they don't use any metal for the length of the stay, which completely eliminates any concern of corrosion. The biggest concern that you need to inspect for is simply chafe. Chafe manifests on synthetic rigging as a fluffy section where the fibers of the strands have broken and now poof out into the wind. A stay that is in good condition will look like a simple straight piece of rope stretched out. A stay with a problem area will have a puffy or fluffy section.

The next issue that can arise with synthetic rigging is if a strand gets cut. Dyneema is made using 12 strands that are braided together; if one of these strands gets cut completely, the strength of the line is now reduced by one-twelfth. If you get a cut strand in a stay, you need to consider replacing the stay or mending it with a new portion spliced in to replace the damaged area.

"I would never install Dyneema rigging on my boat; all it takes is one pissed-off person with a pocketknife to bring my mast down!" I've heard this before, and there are a few major issues with this statement. First, the person who makes this claim has never actually attempted to cut Dyneema because it is really hard to cut, especially when it is pulled tight. Second, steel rigging is just as susceptible to a vengeful person cutting your rigging with bolt cutters and bringing down your mast. The moral of the story is be nice to people and don't worry about vandalism. If the rigging does get a cut in a strand, that section of the stay would just be cut completely and then a mending piece would be spliced in to restore the strength of the stay.

As mentioned, there are several types of Dyneema and then several different manufacturers of the product. Dyneema is a trade name currently owned by Avient Corporation. Dyneema is produced by a few different companies in their respective countries. The two main names you will hear of are DynexDux, produced by Hampidjan in Iceland and sold exclusively through Colligo Marine; and DSM, produced in the Netherlands and sold through a variety of brands such as Marlow, Samson, and New England Ropes.

DSM-produced Dyneema has been used extensively in mining, logging, mooring, and oil and gas drilling, where their fibers are pushed to the absolute limits of comprehensible torture. DSM manufacturers also supply a vast library of research that can answer any and all questions you might develop about the rope, its condition, and its longevity. In their research, they have demonstrated the effects of UV on the fiber over time. Over a ten-year period, UV will gradually cause damage to the outer fibers of the rope. On thin little lines, light will penetrate through the majority of the rope and most of the fibers will suffer from a loss in strength. On thicker lines, only the outer portions of the lines will suffer these effects; the inner fibers will retain their full strength. This is especially apparent on large mooring lines for cargo ships where the line will be several inches in diameter where after years of service, the lines still retain almost all of their original strength. By contrast, based on the findings of a test conducted by DSM, a 3mm line will retain about 15% of its original strength, a 6mm line will retain 25% of its original strength, and an 8mm line will retain about 40% of its original strength. This is after ten years of exposure to the sun! To mitigate this loss, all you need to do is cover the stay with anything that will protect the surface of the fibers from the light. The downside with covering the stay is if a problem were to arise, it is not as easy to quickly inspect. Fuzzy portions or broken strands are much harder to notice if they are covered up.

The decision to cover the stay or leave it exposed is definitely a compromising decision. If you leave it exposed, you can quickly inspect the condition of the line with a simple glance, but at the same time the line is more susceptible to chafe and UV damage. If you cover it with a chafe sleeve, the stay is now protected but also harder to inspect with a quick look. To give a covered stay a proper inspection while still in place, you would have to run your fingers along its entire length to try to feel any depressions where a broken strand has distorted the line's shape inside the cover. The other way is to remove the stay and remove the chafe sleeve to give it a proper visual inspection; however, this is very labor-intensive and probably won't get done as often as it should, which should influence your decision between covered and uncovered Dyneema stays. While retaining less than half of the original strength sounds detrimental, it is important to note that Dyneema rigging is not sized based on breaking strength but rather based on creep—the permanent elongation of the fibers over time when a constant tension is applied. The very weakened Dyneema is still plenty strong to support the load that will be placed on it during sailing. This will be discussed further in chapter 10.

The common types of Dyneema are SK75, SK78, SK99, and DM20. To add to this alphanumerical soup, you can also heat-treat these fibers to change their properties yet again.

RIGGING MATERIALS

The first type of Dyneema to challenge steel was SK75. This fiber is stronger than steel, which is where New England Ropes got their naming system for these lines, STS meaning "stronger than steel." SK75 can work to replace steel in standing rigging but it suffers greatly from creep.

Dyneema is incredibly strong, but as loads approach a percentage of their breaking point for a short period of time, the line will stretch. When the load is removed, the line will return to its original size. This is the situation for control lines where the loads are only applied for short duration and constantly being adjusted, but standing rigging is not a temporary load, and ideally not being constantly adjusted. Standing rigging needs to have a constant amount of tension applied to it to hold the mast in a supported manner. This long duration load will cause the line to stretch at first, but then the molecules in the fibers will move and the elongation will become permanent. Elevated temperatures will accentuate this phenomenon, which means that if your rigging is going to be holding your mast up while you cruise around the Caribbean, you will just be accelerating the creep process on your rigging. As it elongates, it also becomes slack and ineffective. All of this can lead to a poor outcome with synthetic standing rigging if the wrong fibers or wrong size of fibers are used.

The improvement over SK75 was—you guessed it—SK78! This fiber exhibits significantly less creep, which means that it finally became a suitable option for synthetic standing rigging. SK78 is incredibly strong, has low creep, and is priced competitively when compared to other rigging materials. The reason SK75 didn't get completely replaced by SK78 is because manufacturers began heat-treating SK75, which changes the properties of the rope to give even better resistance to creep than SK78 had. New England Ropes marketed heat-set Dyneema as STS-HSR, opening the floodgates for savvy sailors and cruisers on a budget to build their own rigging with very basic tools and techniques. The heat treatment has some interesting effects on the rope: The first and most notable is that the rope becomes incredibly stiff! SK75 and SK78 are very soft and bend incredibly easily, but once treated they become stiff as a rod, which makes them a little tricky to splice if it is your first experience working with Dyneema.

While heat-set Dyneema was gaining in popularity, another version was released called SK99 which has—you know it—even less creep! SK99 never really caught on as much as heat-set SK75 because it was significantly more expensive and was soon overshadowed by the newest and best type of Dyneema on the market at the moment: DM20.

DM20 is said to have no creep, and I can tell you from personal experience, this is a true statement. On a 50-foot (15m) long stay, SK78 will creep about 6 feet (1.8m), heat-set SK75 will creep about 6 inches (15cm), and DM20 will stretch—but not creep—about ⅛ of an inch (3mm). DM20 is a bit more expensive than heat-set

SK75, but the price is well worth it. Working with DM20 is an absolute joy as the rope is soft and easy on your hands. Working the strands into particular positions to carry out the splices is a very easy task to complete.

Steel rigging shows its age in the form of cracks and corrosion, but the SK flavors of Dyneema show their age in the form of creep. Dyneema passes through three life cycles that are characterized by various amounts of creep. In the first phase of the life cycle, Dyneema rigging will creep a lot and quickly. You can tension the rigging in the morning and by lunchtime the rigging will be so slack that you can jump rope with it. The solution is to tighten it again and tension the rigging once more, knowing that it will creep quickly and require another tensioning in short order. This part of the life cycle can be thought of as the toddler phase of your rigging: It has so much potential but instead it just sits there and poops its pants.

Eventually, after a few weeks, the Dyneema will mature and enter the second phase of its life cycle. This phase is what you were expecting when you switched to synthetic rigging. Everything holds well and the creep is barely noticeable. For all intents and purposes, you could easily say that the stay has stopped creeping, but, at a microscopic level, creep is still occurring. For SK78, the creep when loaded to 20% of its breaking strength is roughly 0.03% per year; for a 15m stay, this would be the equivalent to 4.5mm of creep per year. This is a bit excessive but it considers that the stay will be loaded to 20% of its breaking strength and for the entirety of the year. In the next chapter, we will cover how to select the size of Dyneema correctly to minimize creep and make it virtually insignificant. Phase two can be thought of as adulthood: You have strength, health, and the ability to do anything you set your mind to!

Just like with life, the third and final phase is somewhere up ahead, obscured by the veil of time. When synthetic standing rigging enters the last phase of its life cycle, you will know it! The stays that haven't needed to be adjusted for years are suddenly all slack, just like back in the first phase. The incorrect thing to do in this phrase is to tighten the rigging and ignore reality, while the correct thing to do is to start building your new rigging, because it needs to be replaced.

While steel rigging is still the standard on most sailboats, the ease of working with synthetic rigging coupled with the sailor's desire to build something really strong for their own boat has created a niche where synthetic standing rigging can thrive. Now that we have covered all the basics about what rigging is and how the different materials all work together, we can shift our focus to the practical aspect of figuring out what size of standing rigging you need on your own boat and how to build it—be it steel or synthetic.

CHAPTER TEN
BUILDING IT

Knowing how a stay works and how it should be set is only part of rigging; you also need to know how to size the parts accordingly. If you undersize the components, they will break when they should have held; if you oversize the components, they will be more expensive and also become more weight aloft, which is the worst kind of weight to have. For this reason, it is important that you know exactly what size the various components need to be so that you can size everything to work together as one system and therefore as one cohesive unit.

To size your rigging, you need to look at a few different clues. Some of these will cause you to carry out a variety of calculations while others will simply give you a number without any other information. I think it is best to do the math first, and then double-check it with the clues that the boat has provided before you. This is a great way to check your math, and check if the boat was set up properly in the first place.

The most important number for rigging calculations is the boat's RM30, or righting moment at 30°. As the wind builds, the sails fill and the boat heels over. The force on the sails is transmitted via the rigging to the rest of the hull to pull the ballast weight of the keel up towards a horizontal position. As the boat heels over, the force on the rigging increases in the beginning but peaks around 30° of heel. After this point, the sail area begins to diminish and the force on the rigging will be no higher even though the boat continues to heel over. If the boat heels all the way to 90° and the masthead and keel are splashing the surface, the sail will be completely hidden to the wind and while the boat will be as far over as it can be before the dry and wet parts trade places, the force on the rigging will not be much greater than when the boat was heeled to 30°.

There are two ways to go about getting the RM30 value for your boat. One is to carry out an actual test, the other is to use an online calculator that pops out an answer. The actual test is the most reliable method as it gives you the value for your particular boat as you have it set up and loaded. The calculated answer is only as good

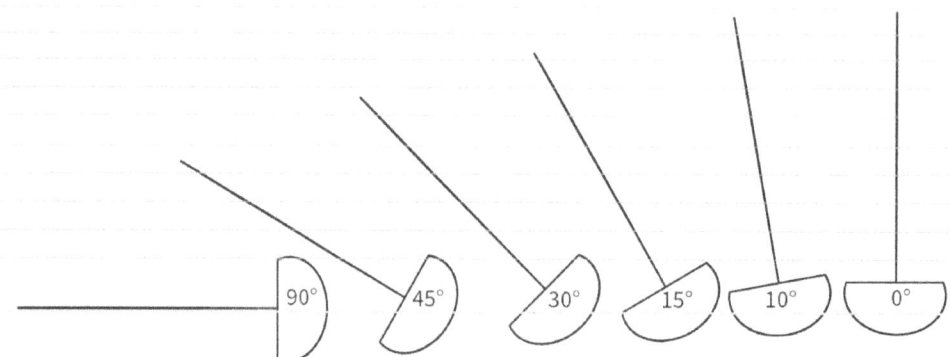

as the numbers you entered and assuming that your boat conforms to the formula that the calculator uses.

To carry out the test, what you need to do is make the boat heel over. and see how much weight it takes to make it heel that amount. The heeling moment force is the force that is trying to make the boat lean over, while the righting moment force is the opposing force created by the ballast in the keel and the flotation of the hull to counteract the heeling moment force. By calculating the force we are imparting on the boat, we can measure the righting moment force for the boat. With your boat resting level and trim, you will observe your clinometer and make sure that it is reading 0°. This means that at this very moment, there is no heeling force and, as a result, no righting force.

In the simplest terms, the test is to figure out how many pounds set how many feet from centerline does it take to make the boat to heel over how many degrees. 1 pound-foot is the force of 1 pound set horizontally to the axis of rotation and 1 foot away. If you move that same 1-pound weight out to 2 feet, it now exerts 2 pound-feet. Your deck is a convenient horizontal surface that is close enough to the axis of rotation of your boat and has plenty of width for you to place weight on to exert a force to cause the boat to heel. If you observe your boat as someone walks around, you will see that the boat will heel over more as they stand on the toerail and heel less as they stand closer to the mast. Having someone with a known weight stand at a set distance from the midline of the boat will cause the boat to heel and this heeling can be read on the clinometer. For the sake of math, you want to try to have the person stand at fixed 1-foot intervals, as well as attempt to make the boat heel over 5°.

If you are working alone, or don't have any friends who can stand still long enough to get a proper measurement, then you can also use a jerry can filled with water: 5 gallons of water weighs 40 pounds and 20L of water weighs 20kg, and you can set them in a line that is parallel to the centerline of the boat. As you carry out the test, it is a good idea to fill out the table below so that you can then average your

BUILDING IT

results to come up with the best answer for your boat's RM30. If you are unable to fill in all the values, you can use basic algebra to figure out the rest of the values on the table. For a small light boat, you will find it easier to get the boat to heel over so you will be able to fill out more of the table. For a large and heavy cruising boat, it will be hard to get it to heel much, so you will be doing a lot of algebra to fill in the voids. I recommend doing the test using weight, even if you prefer the metric system because the SI unit for torque is a Nm, not a kg-m. It is hard to mentally picture 1kg being accelerated 1m per second squared, but it is much easier to simply set a weight statically on the deck and measure the amount of heel.

Angle of Heel	Weight (lbs)	Distance (ft)	Righting Moment (ft-lbs)	Righting Moment per degree (ft-lbs)
0°	—	—		
1°				
5°				
10°				
15°				
20°				
25°				
30°				

While there are no values to enter for 0° of heel, it is always good to have that on the top of your table to make sure that you verified that your boat sits level. If it doesn't then you need to shift weight around inside the boat so that it does sit level before conducting this test.

To fill out the last column of Righting Moment, you simply need to use this formula:

(Distance in feet × Weight in pounds) ÷ Angle of heel

For those with an affinity for the metric system, the same test can be done but the weight will be in *kg* and the distance will be in *m* or *cm*. Once calculated, you can convert kg-m to Nm by using the following conversion: 1kg-m = 9.80665Nm.

You have control over the distance you set the weight, and you also have control over the amount of weight that is set in that spot, so using these two variables, you can control the angle of heel to get 1°, 5°, 10°, 15°, 20°, 25°, and lastly 30°. On a small boat, you can easily get the weight onto the edge of the deck and cause the boat to heel over quite a distance, but on a large cruiser, you might struggle to get the boat to heel over 10°. This is why you measure it at such intervals, first finding the torque needed to heel the boat 1°, then comparing it to the 5° value, which should be either five times the weight or five times the distance. If these two values line up, then you can check again at the 10° mark where it should be ten times the weight or distance as the 1° reading, and double the weight or distance of the 5° reading. Carrying on like this, the righting moment in pound-feet should be the same as you progress through the test. If it doesn't it is best to ignore any outliers and find the average of the most common results. You might be wondering why you need to do all this math if you can simply put a bunch of weight on the toerail and heel the boat over to 30°, this would be, in fact, the force to heel the boat to 30°, which is the value you are looking for! The truth is, the test as it progresses also causes the test to deviate from the ideal. The weight placed on the deck at rest to cause the heeling motion was placed on a horizontal surface that is perpendicular to the floor, so the distance is correct. As the boat heels to 30°, the distance actually gets smaller as the weight swings down and towards the bottom, no longer being as perfect of a test anymore.

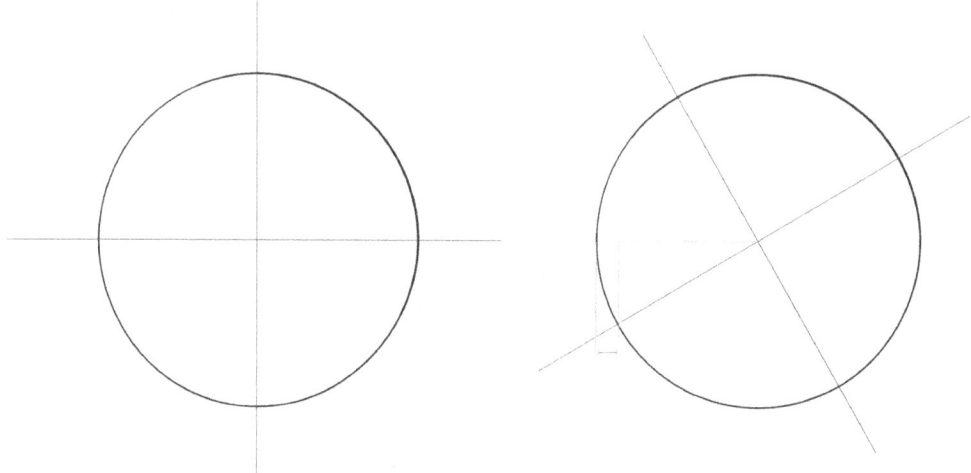

With math, you can account for these errors and use trigonometry to offset these shortcomings, but that starts to become overkill as these tests are in massive units and having the weight a little closer to the midline than expected isn't going to have a catastrophic effect on the outcome of the rigging calculations. It should just serve as a simple reminder that the numbers you will get should be considered the minimum and you should never rig your boat with anything weaker because the test errs on the weaker side.

To show you what the table will look like, let's look at the results from a 1968 Morgan 45-1. This is a heavy boat, displacing about 36,000 pounds when outfitted for cruising and floating on a 32-foot waterline with an 11-foot beam.

BUILDING IT

Angle of Heel	Weight (lbs)	Distance (ft)	Righting Moment (lbs-ft)	Righting Moment per Degree (lbs-ft)
0°	—	—		
1°	520 (13 jerry cans)	4	2,080	2,080
5°	2140 (12 people)	5	10,700	2,200
10°	3890 (22 people)	5.5 (on toerail)	21,395	2,139.5
15°	5840 (32 people)	5.5 (on toerail)	32,120	2,141.3
20°				
25°				
30°				

Fitting more people on the toerail no longer became practical so the test was concluded at 15°. Looking at the numbers from a statistical point of view, we can see that the average is 2140.2. Using this value, we can then complete the table as follows:

Angle of Heel	Weight (lbs)	Distance (ft)	Righting Moment (lbs-ft)	Righting Moment per Degree (lbs-ft)
0°	—	—		
1°	520 (13 jerry cans)	4	2,080	2,080
5°	2140 (12 people)	5	10,700	2,200
10°	3890 (22 people)	5.5 (on toerail)	21,395	2,139.5
15°	5840 (32 people)	5.5 (on toerail)	32,120	2141.3
20°			42,804	2140.2
25°			53,505	2140.2
30°			64,206	2140.2

Test Result: RM30 = 64,206 foot-pounds or 87kNm

This test is best performed on a calm day next to a cement and non-floating pier where you can weigh your friends on a bathroom scale right next to the boat to get an accurate weight of each person. Then you can have them stand around while you stay in the absolute center of the boat watching the clinometer climb to the side as the boat heels over. This test also only needs to be performed once, as once you have the number, you won't need to carry out the test again—unless you forget the result and didn't write it down somewhere.

If you don't have enough friends on hand or the boat is out of the water, then you will not be able to carry out this test and you will be forced to fall back on the much easier option, but also less fun option, of using online calculators to figure out your RM30 value.

The best online calculator can be found on the website of Seldén, a reputable spar manufacturer (https://support.seldenmast.com/en/services/calculators/rm_calculator

.html) or by searching "Seldén RM30 Calculator." To calculate your RM30, you will simply need to enter your boat's draft and beam in millimeters as well as the displacement and ballast in kilograms. Hitting CALCULATE will then spit out your answer in kNm, which is easily converted to pound-feet if you prefer working with the Imperial System. The conversion is 1kNm=737.562 lbs-ft.

Test Boat Results from Calculator: 60,258 pound-feet or 81.7kNm

The practical test results came out to be about 4,000 pound-feet or 5.3kNm higher than the calculated results, which is why it is good to do both, and also good to build in a suitable safety factor to account for these little variances in values. Which one is correct is debatable, as the test has lots of room for error while the calculator doesn't take into account all the variables of the real world. Doing both and going with the higher value of the two is sure to guarantee a safe outcome.

Now that you have your RM30, it is time to put it to use to calculate the rig loads of your boat and in turn determine the minimum wire strength that you will need to have to support your rig loads. The most basic way of thinking of how the RM30 will interact with the rig loads is to realize that the force will be transmitted through all of your rigging and to the boat. If your rigging only had one single shroud, you could then assume that the stay's breaking strength needs to exceed the force of the RM30. As you might have noticed, this is not the case and there are a variety of stays that support the lateral loads on the mast; all of them working together spread out the load and therefore no single stay needs to be strong enough to support the entire force presented in the RM30.

The cap shroud is taking the full load from the top part of the mast as the lowers are taking less of a load and this lesser load is normally shared between the forward and aft lowers. The lower shrouds load is not purely caused by the force of the wind pressure on the sail, as it also has to counteract the compressive force of the spreader. You could approach this problem as an engineer and figure out the actual load on each individual stay, or you could make your life easier and just use the formula below:

RM30 ÷ 3.3 = Practical Rig Load
Cap Shrouds and Intermediate Shrouds: 45% *of Practical Rig Load* ×
2.5 *(Safety Factor)* = *Line Breaking Strength*
Lower Shrouds: 33% *of Practical Rig Load* × 2.5 *(Safety Factor)* =
Line Breaking Strength

BUILDING IT

Using the above formula, you can derive the breaking strength of your wire standing rigging and then determine the necessary size of the rigging wire. If you are interested in synthetic standing rigging, this part of the calculations is still important to you. You won't be using the steel wire for the rigging, but all your associated parts will be sized based on the size of your steel wire requirements. This means that you will need to figure out the right size of steel wire so that you can then size your clevis pins, chainplates, and everything else involved in your standing rigging.

The breaking strength of wire and the associated diameter of the stainless steel wire is as follows:

Breaking Strength (pounds or kg)	Diameter (mm)	Diameter (inches)
300lbs / 136kg	1.2mm	3/64
530lbs / 240kg	1.6mm	1/16
830lbs / 377kg	2mm	5/64
1,200lbs / 545kg	2.4mm	3/32
2,130lbs / 968kg	3.2mm	1/8
2,640lbs / 1,200kg	3.56mm	9/64
3,335lbs / 1,516kg	4.0mm	5/32
4,800lbs / 2,181kg	4.8mm	3/16
6,535lbs / 2,970kg	5.6mm	7/32
8,535lbs / 3,879kg	6.4mm	1/4
10,800lbs / 4,909kg	7.2mm	9/32
13,340lbs / 6,063kg	8mm	5/16
16,885lbs / 7,675kg	9mm	
18,810lbs / 8,550kg	9.5mm	
19,200lbs / 8,727kg		3/8
20,845lbs / 9,475kg	10mm	
25,220lbs / 11,463kg	11mm	
26,145lbs / 11,884kg		7/16
30,015lbs / 13,643kg	12mm	
33,620lbs / 15,281kg		1/2
35,225lbs / 16,011kg	13mm	
40,855lbs / 18,570kg	14mm	
42,620lbs / 19,372kg		9/16
46,900lbs / 21,318kg	15mm	
53,360lbs / 24,255kg	16mm	5/8

Applying our formula to our test boat, we will find that the required strength for the rigging will be:

Tested RM30:
$$RM30 \div 3.3 = Practical\ Rig\ Load$$
$$64206 \div 3.3 = Practical\ Rig\ Load$$
$$19456.36 = Practical\ Rig\ Load$$

Cap Shroud and Intermediate Breaking Strength
45% of Practical Rig Load × 2.5 = *Breaking Strength*
(19456.36 × 0.45) × 2.5 = 21,888.4 *pounds*
Wire Size Based on Breaking Strength: 11mm or $^7/_{16}$ inch wire

Lower Shrouds Breaking Strength
33% of Practical Rig Load × 2.5 = *Breaking Strength*
(19456.36 × 0.33) × 2.5 = 16051.5 *pounds*
Wire Size Based on Breaking Strength: 9mm or ⅜ inch wire

Calculated RM30:
$$60258 \div 3.3 = Practical\ Rig\ Load$$
$$18260 = Practical\ Rig\ Load$$

Cap Shroud and Intermediate Breaking Strength
45% of Practical Rig Load × 2.5 = *Breaking Strength*
(18260 × 0.45) × 2.5 = 20,542.5 *pounds*
Wire Size Based on Breaking Strength: 10mm or ⅜ inch wire (with less of a safety factor)

Lower Shrouds Breaking Strength
33% of Practical Rig Load × 2.5 = *Breaking Strength*
(18260 × 0.33) × 2.5 = 15064.5 *pounds*
Wire Size Based on Breaking Strength: 9mm or ⅜ inch wire

In this situation, the calculated and actual tested values are pretty close to one another and, as a result, the size of wire that is determined to be needed is also pretty similar. The importance of the safety factor is that it grants you a little wiggle room on the size of wire. If the necessary breaking strength of a wire is just under the calculated amount, you will be OK using it, just keep in mind that you have less of a safety factor built into the wire, but it is still above and beyond what is actually needed as a true bare minimum. Going up a size in the wire to comply with the full safety factor will be detrimental to the boat for reasons that will be explained shortly.

BUILDING IT

The biggest clue as to what the naval architect has calculated to be the correct size for the rigging is the size of the clevis pin that is used to connect the various parts of the rigging. The little hole in the top of the chainplate holds more than just the clevis pin, it holds the architect's true vision for the rig and the loads it is anticipated to undergo.

The size of the clevis pin is something that no one will change, partly because they never stop to think about it. They focus on the wire or other parts of the rigging and simply use the clevis pin that fits the hole, without even giving it a second thought. This means that as a boat ages and owners make various modifications to the rigging over time, be it increasing or decreasing the diameter of the wire or changing from wire to synthetic, the clevis pin size will probably remain constant. This pin hole diameter directly correlates to the strength of everything in the system and you can actually see what the designer was thinking.

If a boat has very small clevis pins, then you know that it will be rather tender in the water while a stiff boat of the same size would then be equipped with a much larger clevis pin.

The same holds true when looking at the rigging on a monohull versus a multihull. The monohull rigging doesn't need to be anywhere near as strong as the multihull rigging because the monohull will heel over as the wind builds. The multihull, however, will remain upright and barely heel over, therefore the rigging is going to be forced to withstand a much greater force and needs to be sized accordingly.

If you are unsure of your calculations for the required strength for the rigging based on the RM30, look no further for validation or correction than your chainplate. The top hole for the clevis pin will tell you if you are on the right track or if you need to revisit your mathematics.

Some reasons that your calculations could vary widely from the clevis pin stature are that the boat has been extensively modified and therefore needs to have the clevis pin size adjusted. This would be from a drastic weight reduction aloft that would cause the boat to heel less, or from an increase in ballast, displacement, or draft. All of these factors will cause the boat to stand more upright in the wind, and as a result, cause more strain on the rigging. This is the very reason why it is important to do the math and some form of the test to corroborate the math, verifying that they are within range of each other. If this were not the case, this book would be much shorter and the chart about chainplate sizing, rigging sizing, and clevis pin sizing would be all that you would need to design the rig for any boat. Riggers would ask you "What size is your clevis pin?" You would tell them, and then they would know the size of everything for your boat! This would work in a perfect world where no one changes anything on a boat and uses the boat exactly as the designer envisions, which we all know is not the case.

While discussing modifications to the rigging, the most common modification you will hear people discuss is "upsizing their rigging because they are going to be sailing offshore." This expression makes my skin crawl for several reasons! First, increasing the diameter of the rigging simply makes the entire rig heavier, which will cause the boat to heel sooner and farther. This will decrease the force on the rigging, which can be a good thing, but it will also make you reef earlier as the boat will always be sailing with the rail in the water. The second issue is people increase the wire diameter without increasing the clevis pin diameter. The entire purpose of "making the rigging stronger" is voided by the fact that the clevis pin will still break at the same amount of load!

This makes as much sense as someone putting a gigantic engine in their sailboat with a ton more horsepower thinking that this will let them outrun storms. Adding more power to a displacement hull simply means that you will reach hull speed without pushing the throttle lever as far forward. Once a displacement hull reaches hull speed, adding more thrust will not make it move any faster as hull speed is based on waterline length and the engine's power is not even factored into the equation. If the boat was able to reach hull speed with the old motor, dropping in a bigger motor will not allow the boat to outrun anything better than it did before the swap. All it will do is allow you to burn fuel at a faster rate. If you were curious, the equation for hull speed is:

$$Hull\ Speed = 1.34 \times \sqrt{Waterline\ Length\ (in\ feet)}$$

If your boat has a waterline length of 30 feet:

$$Hull\ Speed = 1.34 \times \sqrt{30}$$

$$Hull\ Speed = 7.34\ knots$$

If the boat has a 20hp or a 200hp motor, the hull speed will still be 7.34 knots and the boat will move no faster than this. 176 nautical miles per day will be your limit, unless the current is helping push you along, and if there is a storm approaching, you will not keep up with a center console boat with a 200hp outboard on the back.

As we return to rigging, it is important to look at why the clevis pin hole should not be changed without proper consideration. The clevis pin was selected by the naval architect based on how they envisioned the boat to be used. If the boat is supposed to be a stiff racer that will fly all of the sails in a gale while up on plane, then the architect will have selected a large clevis pin because they know that the rig loads will be monumental. If the designer envisioned the boat to be a knockabout daysailer that requires the crew to walk out on the keel to bring the mast out of the water when a puff of wind blows through the lake, then the designer would have selected a small clevis pin as the rig loads would be very low.

BUILDING IT

When we look at the clevis pin in this light, we see that the clevis pin is actually a fuse in the rigging, the same way that the electrical system has a breaker panel to protect the rest of the wiring in the boat. Should a wire short out or become overheated due to too much current being drawn on the system, the breaker will trip and protect everything. This protects the wiring from burning and protects the boat from catching on fire. The clevis pin serves the same purpose, but instead of working like a circuit breaker, it works more as a fuse. If the loads are too high, the clevis pin will break and the rigging will separate from the boat. This does lead to the catastrophic failure of the rig, as the mast might come toppling over, but this is actually a good thing! If the rigging becomes overloaded and doesn't break, then that cataclysmic force will be transmitted to the rest of the boat and some other weak link will give way, leading to the same result of losing the rig while also suffering major structural damage!

Picture the SV *Failboat* on a broad reach in a gale, full sail! Imagine if the rigging was infinitely strong and able to withstand these loads. The boat is at hull speed and riding up on the bow wave while the transom is quickly sinking into the stern wave. Then, the captain gets the great idea to jibe by simply turning the helm, letting the sails "do their thing" by slamming over to the other side when the wind wants them too. When the main slams to the other side, the rig loads are going to increase to incomprehensible levels but the rigging won't break because it's infinitely strong. What will happen as a result? The chainplates will rip out and take the bulkheads with them! This *Failboat* might be so lucky as to have the bulkheads tear a hole in the bottom of the boat, which would quickly sink the entire endeavor. Now imagine this same comedy of errors from SV *Failboat* but with real-world rigging that does have limitations and carries out the same maneuver in a gale. The result is predictable: The rigging will break at its weakest point and the mast will fall down, bringing the boat to a quick halt as it loses speed through the water. The boat's hull remains intact and they fall back on their 200hp turbo diesel that they installed to replace their puny 20hp motor that the boat originally came with. The weakest part of the rigging acts as a fuse to break and prevent further structural damage to the boat. Losing the rig is by no means a minor issue, but it will protect the rest of the boat from breaking under these incredible forces that the boat is being subjected to.

When changing the rigging, either erroneously upsizing the wire or converting to synthetic rigging, the clevis pins really should remain the same unless all the engineering is carried out to determine the new size of clevis pin and to re-engineer the entire rig as well as hull to support the additional load that could be placed upon it to break the newly sized clevis pin.

If in your testing of the boat, you find that the RM30 is drastically different from the calculated RM30 based on your boat's designed parameters, and this new RM30 value puts you into a different size of clevis pin, it would be worth careful examination to see if there is anything you can do to bring it back within specification, as this is most commonly caused by the boat being overloaded with "stuff." This overweight situation will cause the boat to sit lower in the water, which will make it sail on lines that it was not designed to sail on, as well as place more load and strain on the boat. For this reason, you are going to be faced with the two options: reducing your boat's weight to bring it back to a safe level or completely rebuild the entire hull and structure to support the new parameters. One of these is obviously easier to do than the other, which is why you should begin unloading your lockers to remove all the "spare parts and good junk" that you carry in the boat and bring the waterline back to where it is supposed to be. Remember, the solution to the boot stripe being underwater isn't to raise the boot stripe!

Breaking Strength (pounds or kg)	Diameter (mm)	Diameter (inches)
2,173lbs	4.78	3/16
3,864lbs	6.35	1/4
6,076lbs	7.95	5/16
8,720lbs	9.53	3/8
11,944lbs	11.13	7/16
15,440lbs	12.70	1/2
24,230lbs	15.88	5/8
34,960lbs	19.05	3/4
47,664lbs	22.23	7/8
58,996lbs	25.40	1

If you are interested in having a little warning to the overload status of your rigging before something just snaps on you, consider using bronze clevis pins instead of stainless steel. They have almost the same ultimate shear strength, so they will break at the same point, but the bronze clevis pin is softer than stainless steel. This means that it will deform when it gets near its breaking point where stainless steel will hold firm until the very end. If you have bronze clevis pins and notice one is bent a bit, you can simply swap that pin out and keep an eye on what maneuver or conditions are causing that pin to bend. This will let you sail in a safer way to prevent future damage on the rest of your boat. Stainless steel pins will not give you this kind of warning, as their yield strength is very close to their ultimate shear strength, so the warning comes right before the end with stainless steel!

Having a basic understanding of the necessary size for the rigging cable, clevis pin, and chainplate, as well as how they all play into each other is important as we move onto the next step where we will go over calculating the size needed for all of these parts. This will let you learn how to calculate and design the size of components needed to safely build the rig in the correct size for your boat and its intended purpose. You can also use all of this information to make sure that the rigging you have on your boat is the correct size! If you are going to be converting to synthetic rigging, this will be the first step in your calculations.

CALCULATING THE SIZE OF THE RIGGING:

Cap Shroud and Intermediate Rigging Strength

$(RM30 \div 3.3) \times 0.45 \times 2.5 = Breaking\ Strength\ for\ the\ Cap\ Shrouds$

Lower Shroud Rigging Strength

$(RM30 \div 3.3) \times 0.33 \times 2.5 = Breaking\ Strength\ for\ the\ Cap\ Shrouds$

Chainplate Strength

$(Breaking\ Strength\ of\ Shroud \div 3) \times 4 = Breaking\ Strength\ for\ Chainplate$

Synthetic Rigging Strength

$(Breaking\ Strength\ of\ Steel\ Stay \times 0.25) \div 0.10 = Breaking\ Strength\ for\ Dyneema$

BUILDING THE RIGGING WITH STEEL

For steel rigging, the wire size is easily calculated with the formula above. Now it is just time to select the correct components to attach the steel wire to your boat. The fastest way to do this is by using a swage fitting, which simply crimps the fitting onto the end of the wire and clamps it in place.

Swage fittings are a favorite of those manufacturing rigging, first because it is quick and easy to do. Simply cut the wire to length, slip the swage fitting over the end of the wire, and crush it down in the swage machine. The second reason they are a favorite is they guarantee repeat business as the process of swaging work hardens the metal and creates stress cracks that will show up sooner than later.

The ideal way to connect wire to a terminator is to use a compression fitting that binds the wire strands over a metal cone that pinches them so tightly inside the metal fitting that they become stuck. There are two main companies that dominate the market and parts for these two companies are readily available worldwide. They are Sta-Lok and Norsman.

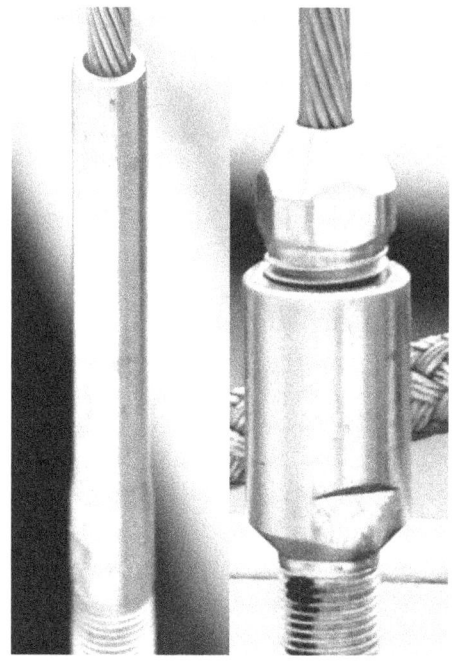

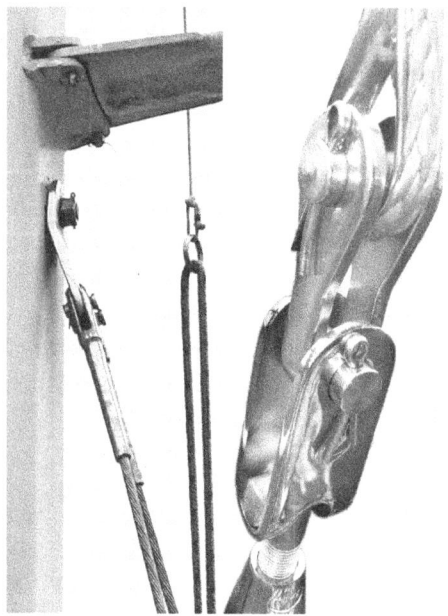

Swage fittings are single use and need to be replaced along with the wire at the end of the service life of the wire rigging, which is ten years or at the first sign of severe degradation. Compression fittings can be reused over and over as the wire is changed out every decade or sooner. Since compression fittings have a much longer expected service life, they are built to withstand these forces with ease by being significantly more robust. Eventually, they will develop small cracks that are only discoverable during a teardown inspection, when all the parts are opened up and cleaned. This usually occurs every decade when the wire is being replaced, or sooner if the wire fails.

If building a rig with steel wire, compression fittings are the ideal way to go. The process of making the rigging is more time-consuming and a little more labor intensive. Compression fittings do not cause a lot of stress on the wire as it enters the fitting like a swage piece would, and the fitting itself is not being work hardened during the manufacturing process. It is important to install toggles at the ends of the fittings to create a fully articulating joint that will reduce the stress on the fittings and reduce the risk of cracks at the eyes of the terminators.

Both Sta-Lok and Norseman offer full lines of components to connect any and all fittings to your rigging, allowing you to connect to your mast with stemball, T-ball, or regular eye fittings. At the bottom, the terminators will end in a screw that will thread into the top of your turnbuckle, or end in an eye, which then attaches to the top jaw of the turnbuckle. However you want to build your rigging, they have a method to make it happen with their parts.

BUILDING THE RIGGING WITH DYNEEMA

If you want to use synthetic rigging, your best option will be to use Dyneema. This fiber is stronger than steel, less dense than water, doesn't get wet, and is resistant to almost all the nasty chemicals that you can think of. The questions that come up are "What size rope should I use to build my rigging?" and "How do I connect the rigging to my boat?"

The first question is honestly the most important, as synthetic rigging is not steel rigging and the same sizing method should not be used. Steel rigging is tightened to a static load of no more than 25% of its breaking strength. This means that if the steel wire has a breaking strength of 10,000 pounds, it will be tightened to no more than 2,500 pounds. You might be wondering what happens if the stay needs to be tightened to 3,000 pounds, and the answer is simple: A larger wire with a breaking strength of 12,000 pounds would be used where 3,000 pounds is 25% of the breaking strength. Each stay of each rig design is tightened to its own particular percentage of breaking strength, but none of them will ever exceed 25% of their breaking strength when they are at rest and the boat is tied up to its mooring.

If you look at the numbers, size for size, Dyneema is much stronger than steel, so to reach this 25% value, you could actually use a much smaller diameter line. This would both save weight and wind resistance! The problem is that Dyneema will creep significantly when subjected to a continuous load that exceeds 15% of its breaking strength.

Stainless Steel Wire	Breaking Strength	Dyneema Size HSR75	Breaking Strength
1/8 inch (3mm)	1,895 lbs	1/8 inch	5,200 lbs
5/32 inch (4mm)	2,965 lbs		
3/16 inch (5mm)	4,265 lbs	3/16 inch	9,475 lbs
7/32 inch	5,810 lbs		
1/4 inch (6mm)	7,585 lbs	1/4 inch	12,385 lbs
9/32 inch (7mm)	9,605 lbs	9/32 inch	18,700 lbs
5/16 inch	11,855 lbs		
3/8 inch	17,500 lbs	3/8 inch	23,600 lbs
7/16 inch (11mm)	24,000 lbs	7/16 inch (11mm)	31,530 lbs
1/2 inch (12mm)	26,900 lbs	1/2 inch (12mm)	43,675 lbs

Creep is different from stretch. When something stretches, it will return to its original length once the force acting upon it is removed. Creep is the *permanent* elongation of the material; and when the force is removed, the length will not return. Dyneema is made of very long chains of polyethylene molecules. These molecules are well connected to each other and during creep, they will actually slide past each other into a new position. When the force is removed, they will stay in their new position. Naturally, if a strong enough force is constantly being applied to cause endless amounts of creep, the molecules would slide past each other until the molecules would run out of length and the fiber would break.

If you can keep the constant load below this threshold, you will not have to deal with creep and the rigging will work just like steel rigging but without all the issues of corrosion, weight aloft, and tiny points of failure hiding in plain sight.

The formula for sizing your synthetic rigging is very simple:

Breaking Strength of Steel Stay $\times 0.25 \div 0.10 =$ *Breaking Strength for Dyneema*

As mentioned before, you need to figure out the breaking strength of a steel stay first as this is going to make the work flow more easily. Since steel stays are not going to exceed 25% of their breaking strength, you can safely assume that the static tension load on the stay will be less than 25% of the breaking strength. This tells you a theoretical realistic load that the stay will be tensioned to in pounds. While synthetic rigging is very strong and a small line could easily support the given load, you want to be sizing the stay based on creep and not based on breaking strength.

The theoretical static load should not exceed 15% of the breaking strength of the Dyneema in order to keep creep down. Therefore, it is best to calculate using 10% of the breaking strength of the Dyneema. This means that when the rigging is tensioned, the Dyneema in the stay will be barely breaking a sweat and this will keep creep to a minimum.

The range I like to use is 5% of the breaking strength for the static load and will have minimal creep. 10% of the breaking strength for the static load will have acceptable creep. 15% of the breaking strength for the static load will have a fair amount of creep but it will still hold up to the task at hand. The reason to even consider the 15% size range is purely because of price. As the diameter of the Dyneema increases, so does its cost and the amount that is consumed in making the splices. Synthetic rigging is a favorite for two completely different markets of sailors: the rich racer and the poor cruiser.

The rich racer wants synthetic rigging because it will reduce weight aloft, allowing the boat to carry more sail without heeling as far, which will make the boat's underwater appendages work better reducing their leeway slippage and improving their control over the helm. The poor cruiser wants synthetic rigging because it can be constructed on the boat with minimal tools or specialized equipment by them, which further reduces the cost of rigging; it also is immune to corrosion and holds up very well to the marine environment. Lastly, it is incredibly easy to inspect, which makes both types of sailors very happy to have this supporting their mast. The rich racer can afford the 5% sizing while the poor cruiser will squeak by on the 15% sizing.

BUILDING IT

Rich Racer

Breaking Strength of Steel Stay × 0.25 ÷ 0.05 = Breaking Strength for Dyneema

Standard Sizing

Breaking Strength of Steel Stay × 0.25 ÷ 0.10 = Breaking Strength for Dyneema

Poor Cruiser

Breaking Strength of Steel Stay × 0.25 ÷ 0.15 = Breaking Strength for Dyneema

Diameter (mm)	Steel Wire 316	Heat Set SK75	SK78	DM20
1/8 inch 3mm	1,895 lbs	5,200 lbs 2,363kg	2,800 lbs 1,273kg	2,970 lbs 1,350kg
3/16 inch 5mm	4,265 lbs	9,475 lbs 4,306kg	6,050 lbs 2,750kg	6,314 lbs 2,870kg
1/4 inch 6mm	7,585 lbs	12,385 lbs 5,630kg	9,700 lbs 4,409kg	9,042 lbs 4,110kg
9/32 inch 7mm	9,605 lbs	18,700 lbs 8,500kg	14,500 lbs 6,590kg	14,828 lbs 6,740kg
11/32 inch 9mm	15,008 lbs	23,600 lbs 10,727kg	18,800 lbs 8,545kg	21,318 lbs 9,690kg
7/16 inch 11mm	22,418 lbs	31,530 lbs 14,331kg	24,000 lbs 10,909kg	29,920 lbs 13,600kg
1/2 inch 12mm	26,680 lbs	43,675 lbs 19,852kg	34,300 lbs 15,590kg	
9/16 inch 14mm	36,316 lbs	55,770 lbs 25,350kg	42,000 lbs	
5/8 inch 16mm	45,194 lbs	65,950 lbs 29,997kg	58,000 lbs	
3/4 inch 18mm	51,717 lbs	86,350 lbs 39,250kg	67,000 lbs	
25/32 inch 20mm		103,650 lbs 47,113kg		
7/8 inch 22mm		112,760 lbs 51,254kg	97,500 lbs	
1 inch 24mm		130,900 lbs 59,500kg		

There are three types of Dyneema that people are using for their rigging: Heat Set SK75, SK78, and DM20. Two of them are manufactured for the intention of being used as rigging, while SK78 simply seems to be having good results by sailors willing to experiment a bit.

Before looking at how to build a synthetic rig, it is important that we discuss which boats should and should not have synthetic rigging. In the most basic sense, it depends on what is asked of the rigging and what the intended use of the boat will be.

The ideal candidate for synthetic rigging is a boat with a single set of in-line spreaders with a standard mainsail that gets raised and lowered. This rig is under the least amount of tension and is the least sensitive to rig tensioning. As the temperatures vary, and the Dyneema expands and contracts with the temperatures (more on this later), the rig will remain in column and the crew will sail happily.

The least ideal rig is a linked rig with multiple sets of aft swept spreaders and in-mast furling. These rigs require a specific amount of rig tension and if they are too tight or too loose, the mast will come out of tune and can even collapse!

WHERE ARE YOU SAILING?

Are you sailing year-round or only in one particular season? Do you have *real* seasons where you are going to be sailing? These are important questions to ask, as they can become major problems if you are going to be working against them. Stainless steel has a coefficient of thermal expansion that is almost identical to aluminum, so as the temperature swings, the aluminum mast and steel wire will expand and contract at almost the same rate.

$$\text{Dyneema: } -12 \times 10^{-6} m/K$$
$$\text{316 Stainless Steel: } 15 \times 10^{-6} m/K$$
$$\text{Aluminum: } 23.1 \times 10^{-6} m/K$$

Over a 15-meter length, if the air cools down 30°C (54°F), from summer to winter, the materials will change in lengths in the following ways:

Dyneema will expand 5.4mm.
Steel will contract 6.75mm.
Aluminum will contract 10.4mm.

If you have steel rigging with an aluminum spar, the difference in lengths will only amount to around 3.65mm since they are both contracting together, which isn't a very noticeable value in the tension of the rigging. The same aluminum mast with Dyneema rigging will be much greater as one is expanding as the other is contracting. The same temperature swing will result in the rigging becoming approximately 15.8mm too long, which will result in the rigging going completely slack and the mast will become unsupported by the slack rig tensions.

Having a keel-stepped rig that is less dependent on static rig tension over winter or a deck-stepped rig that will only be sailed at a particular time of year and the mast is removed for winter storage will be the only cases where this temperature variation wouldn't become a problem. If you are planning on sailing year-round in an area with strong seasonal changes or cruising from the tropics to the arctic in one direct passage, then synthetic rigging would not be the wisest of choices for you to make.

As a blanket statement, if you have in-mast furling, you really need to stick with steel rigging as the mast needs to remain absolutely straight with no bend or rake for the sail to safely furl or unfurl. Only sailboats with a traditional mainsail that are raised and lowered should consider converting to synthetic rigging.

MASTHEAD RIG WITH ONE SET OF IN-LINE SPREADERS

This rig is the least sensitive to rig tension as it can survive with the lowest rig tension and still maintain itself upright. At rest, the mast is simply held in place by the rigging, deriving its shape from the metal it is made of. Keel-stepped versions are even better than deck-stepped versions with synthetic if you will be sailing in areas with greater temperature swings. Dyneema rigging becomes longer as it cools, which is compounded by the shrinkage of the aluminum spar as the temperature drops. The result is very slack rigging! If you sail only in the tropics, where the temperature is always nice and hot, then your rigging will remain nice and tight. If you sail in more temperate climates where summer and winter are real seasons, then your rig tension will vary wildly as the seasons change. In the dead of winter, a keel-stepped mast will still remain supported by the mast step and mast/deck interface. On a deck-stepped mast, the mast may lean over a bit during a cold winter, which could damage the mast heel as it leans over on the mast step.

The biggest advantage of this rig is as the temperatures change, the entire rig will change in lengths and tensions the same. The lowers will expand half as much as the cap shroud, as they have half the distance to travel to reach the mast. When you load the sails, the mast will lean over a smidge until the stays become tight and the mast will still hold true, just that it might be leaning over a little more than usual. It is prudent to keep the sails lower on the mast via reefing if the temperatures are very cold to lessen the burden on the spar as it would need to lean over a bit before the rigging starts to help support the load. It is also prudent to never crash jibe on a cold day as the rigging will be slackened and the shock loads will be monumental!

These rigs are the easiest to set up and tune as well, as the mast will either be leaning a bit much, requiring the cap shrouds to be tighter, or bowing in the middle, requiring the lowers to be adjusted. There will never be an S bend requiring analysis to determine if it's the lower or the intermediate that needs correction. For a beginner setting up their own synthetic rigging, this is the best rig design to work on.

MASTHEAD RIG WITH MULTIPLE SETS OF IN-LINE SPREADERS

This setup is a good candidate for synthetic rigging only if you have a continuous rig. Adjusting the intermediates can be tedious, and is best fine-tuned while under sail with the rig loaded up. This lets you see the direct results of your adjustment with real-world results. If you have a continuous rig, the adjustments are made at the deck, where a linked rig would require you to go aloft while undersail to adjust the high side turnbuckle on the spreader tip. If you have a triple spreader rig, then you will need to climb the mast up to the first and second set of spreaders! The worst part is that you can't really sight the mast from the spreader to see if your adjustment is working like you expect it to. Suddenly you need to have a spotter at the deck calling out orders for you to carry out while dangling from the windward spreader tip in a stiff breeze. I don't know about you but that just sounds miserable to me!

The cap shrouds need to be tensioned to 15% of the breaking strength of a steel stay that is appropriate for that rig, which will be much less than 15% of the breaking strength of the Dyneema that will be in use. The intermediate and lowers simply need to have enough tension on them to counter the compressive force of the spreader pushing inward on the mast. This is where adjustments can get a bit hairy as a slack lower will cause the mast to bend in an S shape with the lower bow being to the leeward side, and the upper bow being to the windward side. A slack intermediate will cause the opposite S shape, with the lower bending towards the windward side and the upper section bending to leeward. Sometimes, people get confused and think that the lower is too tight and then get frustrated when the bend doesn't resolve as they expected by loosening the lower when in reality they should have tightened the intermediate. This frustration would only be compounded by a linked rig where the failure to get the results was magnified by the complexity getting to the turnbuckle.

Multi-spreader rigs can be done with synthetic rigging, but you really should be proficient in tuning your mast, that way you can then tune it properly and accurately while dealing with the added layer of confusion that can come from temperature changes and creep.

FRACTIONAL RIG WITH ONE OR MULTIPLE IN-LINE SPREADERS

Fractional rigs are popular because they allow you greater control over the tension of your headstay. This directly affects the amount of headstay sag and, as a result, the boat's windward performance. These boats are favored by racers or sailors that just want to sail as fast as they can. The conversion to synthetic rigging will be well rewarded as this will greatly reduce your weight aloft and allow you to carry more sail as the breeze builds. Sailing a flatter boat also means that the keel will remain more vertical, which will reduce your leeway slip. The cap shrouds and backstay should be tensioned to 15% of the breaking strength of a steel stay, so the loads on the Dyneema are not going to be that high, relatively speaking. The backstay should have an adjuster in it so that you can better tune the headstay tension based on the temperature. If it's cooler, adding a little bit of backstay tension will take any slack out of the headstay and bring your headstay back up to the target tension for the best sailing. A single spreader setup will be easiest to set up and tune, but a multi-spreader rig can also work as long as it has a continuous rig and all the stays terminate at the deck.

MASTHEAD RIG WITH ONE OR MULTIPLE SETS OF AFT SWEPT SPREADERS

Aft swept spreaders are an excellent invention that uses the tension on the cap shroud to push the middle section of the mast forward, creating a better bend and flattening the mainsail as well as aiding in the aft pull of the masthead, sharing the load of the backstay. In other words, as the wind builds, an aft swept spreader will actually help trim your mainsail for the increasing wind conditions automatically, and within reason as a flatter mainsail will depower itself, but not as well as reefing would.

Naturally, this wonderful system only works if the rig tension is set properly. Imagine a clock, perfectly assembled but the mainspring doesn't have the right amount of tension, suddenly the clock won't keep time properly and the clock runs slow. Steel rigging is able to keep the tension as the temperatures vary because it has a relatively similar coefficient of thermal expansion to aluminum. This means that as the temperatures rise and fall, the mast and rigging will expand and contract together while still holding a relatively similar tune. This is how sailors with aft swept spreaders can go sailing in the middle of summer and then also be in the Parade of Lights during the winter without having touched their rigging.

Dyneema expands as it cools and contracts as it warms, which is opposite to what the aluminum mast would do. As a result, the rigging would not hold its tension as the seasons changed and the wonderful functionality of the aft swept spreaders would be lost. Synthetic rigging could potentially work for a racing boat that needs the performance benefits of synthetic rigging, but they would also need to have their rigging tuned before each sail to make sure that the tensions are correct for the day's temperature. This is honestly too much effort and really should be avoided, but if you must convert to synthetic, it is important that you know what you are getting yourself in to.

The cap shrouds need to be tensioned to 25% of the steel's breaking strength and the backstay to 15% of the steel's breaking strength. These tensions are very important to maintain as they keep the mast arching backward, which is the way it is designed to bend. If these tensions become loosened, the mast can suffer an inversion, which is a catastrophic failure for the spar and will necessitate replacement. As you can imagine, that's a really bad thing, which is why masthead rigs with aft swept spreaders are not good candidates for synthetic rigging.

FRACTIONAL RIG WITH ONE OR MULTIPLE SETS OF AFT SWEPT SPREADERS

This rig should really stay with steel rigging. The tolerance for anything less than an exact tension amount is nonexistent and the demands for perfection are absolute. The cap shrouds need to be tensioned to 25% of the steel's breaking strength and the backstay needs to be tensioned to 20% of the steel's breaking strength. This rig needs to be under constant tension, and if you are sailing near a cold front, the rapid cooling of the air can throw your tension out the porthole and your rig will be in danger of collapsing due to becoming out of spec and unsupported. These rigs are incredible and perform superbly to windward; but just like a sports car needs to have a specific octane rating for its fuel, these rigs need a specific tension in the rigging. You bought the boat for the performance characteristics and you know that it needs to have steel rigging to perform like it should!

This is a very short section in the book because these rigs really shouldn't be set up with Dyneema, as the temperature sensitivity will be unacceptable.

Examples:

Knot Normal

LOA: 30 feet
Beam: 9 feet
Draft: 4.5 feet
Displacement: 9,000 lbs
Ballast: 3,000 lbs
Single in-line spreader masthead rig
Poor Cruiser

RM30: 15.5kNm or 11,432 lb-ft

Cap Shroud and Intermediate Rigging Strength
$(RM30 \div 3.3) \times 0.45 \times 2.5 = $ *Breaking Strength for the Cap Shrouds*
3,897 pounds = Breaking Strength for the Cap Shrouds

Lower Shroud Rigging Strength
$(RM30 \div 3.3) \times 0.33 \times 2.5 = $ *Breaking Strength for the Cap Shrouds*
2,858 pounds = Breaking Strength for the Lowers

Synthetic Rigging Strength
Breaking Strength of Steel Stay × 0.25 ÷ 0.15 = *Breaking Strength for Dyneema*

6,495 pounds for the Cap Shroud = 5mm Heat Set SK75 Dyneema (9,475 pounds breaking strength)

4763 pounds for the Lower Shrouds = 3mm Heat Set SK75 Dyneema (5,200 pounds breaking strength)

Backstay, Headstay, Cap Shroud: 5mm Heat Set SK75 Dyneema
Lower Shrouds: 3mm Heat Set SK75 Dyneema

Sometimes, you can actually save money by buying a full spool and getting a bulk discount for one size and using that one size for the entire rig. In this case, you would buy a full spool of 5mm and know that the lowers are oversized.

Seas the Day

LOA: 50 feet
Beam: 14 feet
Draft: 6 feet
Displacement: 25,000 lbs
Ballast: 10,000 lbs
Double in-line spreader continuous masthead rig
Standard Sizing

RM30: 71kNm or 52,367 lb-ft

Cap Shroud and Intermediate Rigging Strength
(*RM30* ÷ 3.3) × 0.45 × 2.5 = *Breaking Strength for the Cap Shrouds*

17,852 pounds = Breaking Strength for the Cap Shrouds

Lower Shroud Rigging Strength
(*RM30* ÷ 3.3) × 0.33 × 2.5 = *Breaking Strength for the Cap Shrouds*

13,092 pounds = Breaking Strength for the Lowers

Synthetic Rigging Strength
Breaking Strength of Steel Stay × 0.25 ÷ 0.10 = *Breaking Strength for Dyneema*

44,630 pounds for the Cap Shroud = 14mm Heat Set SK75 Dyneema (55,770 pounds breaking strength)

32,730 pounds for the Lower Shrouds = 12mm Heat Set SK75 Dyneema (43,675 pounds breaking strength)

Delta Foxtrot Lima

LOA: 40 feet
Beam: 13.2 feet
Draft: 8.5 feet
Displacement: 10,902 lbs
Ballast: 4,960 lbs
Triple in-line spreader continuous fractional rig
Rich Racer

RM30: 34.2kNm or 25,224.63 lb-ft

Cap Shroud and Intermediate Rigging Strength

$(RM30 \div 3.3) \times 0.45 \times 2.5 =$ *Breaking Strength for the Cap Shrouds*
8,599 pounds = Breaking Strength for the Cap Shrouds

Lower Shroud Rigging Strength

$(RM30 \div 3.3) \times 0.33 \times 2.5 =$ *Breaking Strength for the Cap Shrouds*
6,306 pounds = Breaking Strength for the Lowers

Synthetic Rigging Strength

Breaking Strength of Steel Stay $\times 0.25 \div 0.05 =$ *Breaking Strength for Dyneema*
42,995 pounds for the cap shroud = 12mm heat-set SK75 Dyneema (43,675 pounds breaking strength)
31,530 pounds for the lower shrouds = 11mm heat-set SK75 Dyneema (31,530 pounds breaking strength)

With all of these examples, you might have noticed that the formula to calculate the size for the synthetic rigging assumes that the steel stay was loaded to 25% of its breaking strength. As you know, that only happens on boats with aft swept spreaders; all other rigs with in-line spreaders only reach loads of 15%. The reason for the added 10% is to further the safety factor. The more oversized the Dyneema is for the task, the less creep it will experience. If you size the rigging for the Poor Cruiser at 15% for an in-line spreader instead of 25%, you will end up with a thinner line and way more creep than would be acceptable. By using the value of 25% of the breaking strength of steel, and then further adjusting based on creep toleration, being 5% for minimal creep, 10% for standard sizing, and 15% for the most budget oriented cruiser, you will end up with a predictable and stable result that you can with confidence rely upon.

CONSTRUCTING THE RIGGING

Knowing what size of parts to select to build the rigging is very important, but now you have a few spools of heat-set SK75 or DM20 sitting on the floor of your workspace. What do you do with them?

The rules for building a synthetic rig are very simple, and if you follow them to a tee, your work will be strong and reliable.

The Rules

1. Always measure the line before you open the weave to make a splice.

2. The bury amount is 72× the line diameter in millimeters at a minimum.

3. Chafe covers are useful for areas that will be subject to chafe; slide them on before you splice and make their cut length a little longer than the length of the stay.

4. Stretch the stay after you splice it.

The reason you need to measure before you open the weave for a splice is because once you open the weave, the fibers in the line will change angle and the whole line will appear shorter. This is called constructional shrinkage, and is removed by pulling hard on the line to reset the fibers back in their original orientation. As the length that the line gains after the splice is called constructional stretch, and is nothing more than the fibers moving back to their original positions. The moment you introduce constructional shrinkage, you lose your frame of reference for the length of the rope you are working with and if you are measuring after your first splice, the length of the rope at the end is anyone's guess. This becomes a major problem when you are building a headstay that will attach to only a turnbuckle. In this situation, you need to build the stay to an exact length as the turnbuckle only has a small amount of throw before it becomes two blocked. Always measure the stay by wrapping the rope around the thimbles you will use at the length you want it to be. If you are using heat-set SK75, the creep that will occur later will be about equivalent to any shrinkage that might occur from the weave opening up to hold the buried portion of the tail for the eye splice.

The bury is the part of the splice that takes the load so that the eye itself doesn't have to. If the bury is too short, the fibers will slip and the Mobius Brummel eye will take the load. The bends of the splice are weaker than the straight sections of the stay and will become the point of failure. Rigging needs to be as strong as possible, doing the minimum is just not acceptable. The minimum bury is 72× the diameter of the line in millimeters. This means that if you have a 5mm line, you need to bury 360mm of line. You also need to taper the bury so that the transition from single braid to single braid with a buried section inside happens smoothly as any sharp changes will cause more stress and will be the future point of failure of the stay. The full thickness bury is the part that holds the best, as the tapered portion is thinner and doesn't hold as well. For standing rigging, bury double the minimum amount and be confident in the strength of your splices. The taper will occur on the last 72× portion while the first 72× portion is untouched. This will provide plenty of space for a very gradual taper while still providing plenty of line for the splice to be supported. This might sound wasteful, but you will only be using a few extra feet of line for the bury and that is cheap insurance when you think about the cost of a splice failure!

Dyneema Size	Running Rigging Bury (72x)	Standing Rigging Bury (144x)
1/8 inch 3mm	8.5 inches 216mm	17 inches 432mm
3/16 inch 5mm	14.2 inches 360mm	28.4 inches 720mm
1/4 inch 6mm	17 inches 432mm	34 inches 864mm
9/32 inch 7mm	19.9 inches 504mm	39.7 inches 1008mm
11/32 inch 9mm	25.5 inches 648mm	51 inches 1296mm
7/16 inch 11mm	31.2 inches 792mm	62.4 inches 1584mm
1/2 inch 12mm	34 inches 864mm	68 inches 1728mm
9/16 inch 14mm	39.7 inches 1008mm	79.4 inches 2016mm
5/8 inch 16mm	45.4 inches 1152mm	90.7 inches 2304mm
3/4 inch 18mm	51 inches 1296mm	102 inches 2592mm
25/32 inch or 20mm	56.7 inches 1440mm	113.4 inches 2880mm
7/8 inch or 22mm	62.4 inches 1584mm	124.7 inches 3168mm
1 inch or 24mm	68 inches 1728mm	136 inches 3456mm

While synthetic rigging is immune to rust and other forms of corrosion, it is very susceptible to chafe. If sheets, hanks, or even the sail rubs repeatedly on the rigging, the rigging will develop chafe and gradually grind through the thin fibers. Upon inspection, you will see a fuzzy section in the stay and this is your clue to the damage that your rigging has suffered. Upon closer inspection you can see how bad the damage is and how many fibers appear to be affected by the chafe. If the damage is too great, the stay will become compromised and should be replaced. If the cost is too great to replace the entire stay, then the damaged section should be cut out and a mending splice carried out with end-to-end splices above and below the damaged area to attach in the new length of Dyneema. All of this could be avoided by simply using a chafe guard of some form. A chafe sleeve can be slid over the stay and sewn into place once the stay is set up and its position established. For areas requiring greater protection, service can be applied to the section of the stay, which will make that part absolutely bomb proof. It is recommended to service the part of the stay that goes over the spreader tip as this part needs to be well protected. The added friction from the service over the stay also comes in handy when seizing the stay to the spreader tip once the rig is set up and settled in.

Lastly, always stretch your stays out before you install them. This will save you loads of time as most of the constructional stretch as well as initial creep can be taken out at this time. My method of choice is to attach the stay to the tow hitch of an old pickup truck that weighs around three tons with the other end secured to a very strong tree at the top of a hill. Simply drive away from the tree with your speed not exceeding 5 mph or 8km/h until the truck comes to a jarring halt; hence the need to keep the speed low and a seatbelt. Repeating this until the truck doesn't move any farther forward will get the constructional stretch out of the splices but the shock loading won't do much for the creep.

Leaving the truck in neutral with the rope holding it via the tree will cause the stay to be held under a static load. Marking the sidewall of the tires at bottom dead center is helpful as it lets you see the progress in the creep removal. After a few hours, inspect the chalk line and you will see that the truck has rolled forward a bit and the chalk line is rotated away from the position of the bottom dead center. If the line doesn't creep at all, try using a steeper hill or a heavier truck. After a few days, the truck should stop inching forward and you can assume that the creep has been mostly removed. Now when you install the stay on the boat, most of the annoying part when the stay needs to settle in has been taken care of.

When it comes to splicing the eyes at the ends of the stay to attach your rigging, you are faced with two options: the long bury splice and the Mobius Brummel eye splice. The long bury splice is by far the simpler of the two splices, as you simply make an eye and slip the tail back into the line and bury it a really long distance, hence the name. The other splice involves a little bit of wizardry to create, but the end result is a splice with some wonderful qualities.

At first glance, the long bury splice will seem like the way to go as the process of making the splice is more straightforward and the splice itself retains more of the

line's original strength. This means that the end result is a stay that is as strong as possible! Why on earth would we even entertain the notion of a more complicated splice instead? While the ultimate strength of the long bury splice is greater, the splice has a fatal flaw: It can come undone easily. When the stay is loaded, the line will crush down on the buried section just like a Chinese finger trap. The forces and loads are evenly distributed over the entire length of the bury and the splice will hold firm. As soon as the tension is released, the grip from the splice is lost and the buried portion can slip right out.

While sailing, the windward stays will be under load and will hold firm, but the leeward stays will be slack; the wiggles and jiggling of the leeward shrouds in the wind can cause the buried portion to quietly work their way out. When you go to tack, the load will shift to these stays but the strength will be gone as the splice has fallen apart! To remedy this, a locking stitch can be sewn into the splice to secure the buried tail when tension is not present, but this means that you are trusting your entire rig to a small bit of thread. Some might think this is acceptable, but I personally do not. We will cover how to carry out both splices and if you feel like living dangerously, you can splice your rigging using the long bury eye splice. If you are more of a belt-and-suspenders kind of person, then the Mobius Brummel eye splice will let you forget about your rigging and focus on sailing instead.

The Mobius Brummel eye splice forms a lock right between the eye and the buried portion. This lock induces a few extra turns that will weaken the line a little bit, which might sound like a good reason to turn away from this splice until you look at the numbers. The long bury eye splice retains almost all of the original strength of the line, while the Mobius Brummel eye splice only retains 90–95% of the strength of the line.

Imagine that you have determined that you need to use ⅜ inch wire rigging with a breaking strength of just over 17,000 pounds. Using the formula:

$$17{,}000 \; pounds \times 0.25 \div 0.15 = 28{,}333 \; pounds$$

The stay you are replacing had a breaking strength of 17,000 pounds and you are replacing it with something that has a breaking strength in the area of 28,000 pounds. If this is weakened by 10%, it still has a breaking strength of 25,500 pounds, a full 8,500 pounds above the old stay's breaking strength.

My thoughts on the ultimate breaking strength of an eye splice being compromised by using the Mobius Brummel over the long bury are that close to the mark, you should simply increase the size of the line you are working with so that it is strong enough to support the loads required. Dyneema is so lightweight and strong for its size that there really is no reason to use any line that ever approaches its breaking strength.

LONG BURY EYE SPLICE

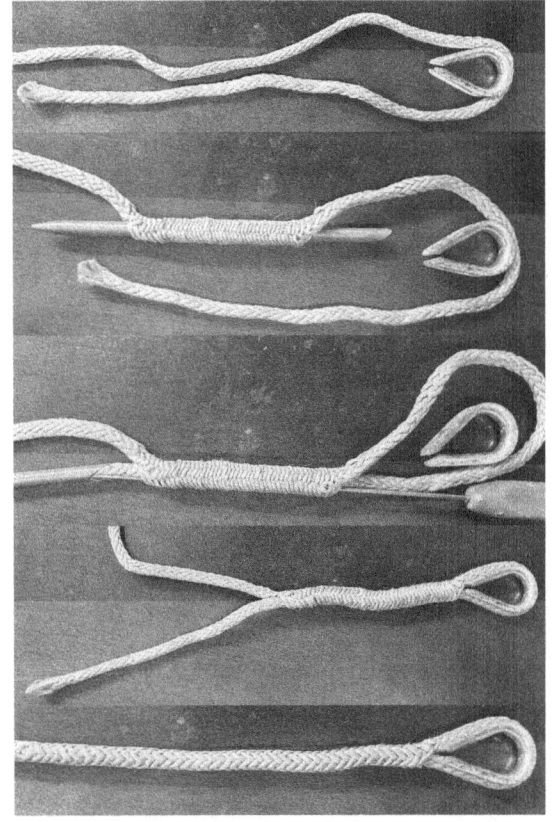

The long bury eye splice is the best first splice to learn as it is the easiest introduction to Dyneema splicing. Looking at the pictures at right, you can see how the process is quite straightforward. The line is laid over the thimble and then buried into the standing part of the line. The only part that really takes any thought is the amount of tail that needs to be buried, which is 72 times the diameter of the line in millimeters. Referencing the chart above, you can easily find the appropriate amount of tail that you need to bury to get the job done securely.

The goal of the taper is to ensure that the transition from "standing part & tail" to "just standing part" is as smooth and gradual as possible. Dyneema is incredibly strong as long as the fibers are running in a straight line. The fibers do not like to be bent, which is why eye splices need to have a thimble in them to keep the eye open. The minimum bend radius is 3x the rope's diameter, and the distance from the throat of the splice to the end of the eye should be 3x the diameter of the size of the eye.

In other words, if you have a 5mm line, the minimum size of the thimble needs to be 15mm wide, and the throat needs to be 45mm away from the end of the eye. This will ensure that the rope is not being bent too tightly and that the throat is not pulled apart or torn, keeping all the fibers in a much happier arrangement, which will lead to a longer service life.

Having any sharp turns in the fiber will lead to the fiber breaking and that becoming the point of failure. If the tail is buried and then comes to an abrupt end, the fibers right at the transition from "standing part & tail" to "no tail" will become stressed and break!

Taper is easy to carry out mathematically, as the rope is made out of 12 strands, all you need to do is count how many pairs of weaves you have pulled out of the standing part and then divide it by 11, as you don't need to trim the twelfth strand for that would just make the tail shorter. If your tail is showing a hundred strands,

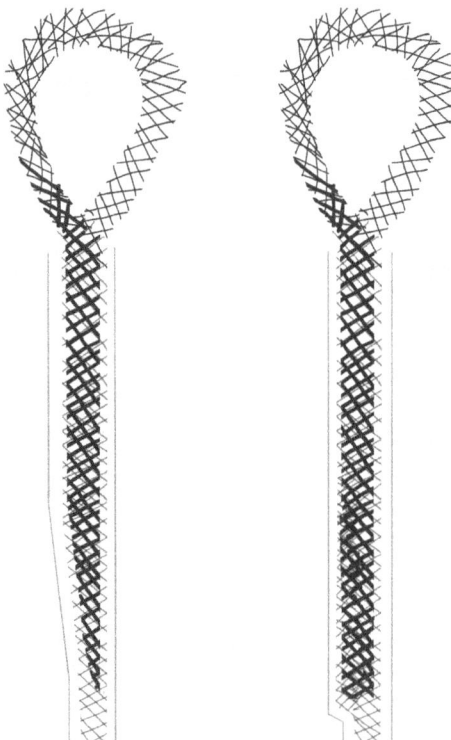

simply take 100 ÷ 11 = 9. This means that every ninth strand needs to be pulled out and then cut off.

Once the splice is completed and the tail has been milked back into the standing part, it is technically completed, but as mentioned previously it can slip apart if there is not a steady load applied. To prevent this issue, you need to sew in a locking stitch; my favorite thread to use is the longest cutting from the tapering process, as it will color-match perfectly and blend right in with the line that it came from.

With the long bury eye splice covered, we can now move on to a more complex splice known as the Mobius Brummel eye splice, which locks on its own and is therefore more secure for use in any application, especially for your standing rigging.

MOBIUS BRUMMEL EYE SPLICE

The easiest way to make a Mobius Brummel is to have both tails free, pass the first tail through the second tail, then the second tail through the first tail. The end result is a solid splice that can't slide apart because they lock each other.

If you don't have access to both tails, there is yet another method that is more complicated but will enable you to make the splice with only one free tail. First, you need to figure out your measurements, as you don't want to redo the splice because you made it too short. The first measurement you need to establish is the length of the buried tail, this needs to be 72 times the diameter of the line in mm. If you are working with 5mm Dyneema, then your buried tail needs to be 360mm long, or 14⅛ inches long. The second measurement is the size of the eye. For let's plan a 3-inch eye, which means we need 6 inches for the loop.

To make the splice, simply make two loops at the end of the line. The span between the loops will be the length of the eye, so if you want to make a 3-inch eye, you need to leave 6 inches between the loops. The loops are easy to make by starting with a straight piece at the tail and simply pushing the fid with the end of the tail through the point of the first loop, pulling the line through but leaving a small loop present. Then pass the fid and end of the tail through the second point as making a second loop. The goal is to have the standing part come up to a loop where the line comes out to the left, then through another loop where the tail comes out to the bottom.

Now pull the second loop all the way until the loop disappears and the line actually twists on itself. At this point, you have preloaded the twist, which you will now take out. Using two fids shoved into the twisted hole, you can work them like a grandma sitting on a train working away with her knitting needles, spreading the sides of the hole to make it bigger and easier for the next step. Simply push the first loop through the enlarged hole and then slide the hole down over the loop. When the second hold makes it past the first crossover, you will see that the twist will flip back to normal and the fibers will run straight. Giving the eye a pull should demonstrate that the splice is locked. If for some reason you can pull the splice apart by pulling on the eyes, then you did something wrong and it didn't work out too well for you. Take the splice apart and try it again until you can show that the splice is locked and secured.

Now for the easy part where you simply bury the long tail, pull it out past the point where it will end up, and taper the end. Taper is all math at this point. Dyneema is made of 12 strands, simply count how many fibers are sticking out from the exposed tail where it exits the standing part to the end of the tail and divide the number by 11, ignoring any decimals. In this example where there is too short of a tail, there were 43 fibers present. 43/11 = 3.9, so 3. This means that each third fiber gets pulled out and then cut off. This will give a uniform reduction to the tail and produce a nice gradual taper from start to finish. With the taper finished, simply milk the tail back into the standing part and this splice is rock solid. Be sure to remove as much constructional stretch and creep from the spliced line as you can before setting it up for standing rigging, but if you are using this for your running rigging, the first time you load it up will pull any constructional stretch out of it.

BUILDING IT

> Specific knots and splices are best explored via the internet where interactive guides exist that can take you step by step through the process. Rather than filling up page after page with a tutorial on how to do them, I recommend checking out our YouTube channel, @Rigging Doctor, or my website, www.RiggingDoctor.com, for step-by-step instructions.

TOGGLES

Connecting the eye splice at the top of the stay is easy if the mast uses regular tangs that employ clevis pins. In these cases, the eye can be connected directly to the tang by having the clevis pin pass through the eye of the splice, with the thimble in place to make sure that the bend radius is respected. If the eye splice looks like it might chafe, their orientation can be rotated 90° by the use of a toggle.

Rigging toggles are simply connectors placed between the stay and the mast tang. They are typically considered an afterthought, since they add between 1.5 inches and 2 inches of length to the stay. Typically, if a stay comes out a bit short, toggles are used to bridge the gap and make it reach. Because of this, toggles have taken on somewhat of a bad image; broadcasting to the world that the stay was measured a bit too short!

The reality is toggles should be used at the mast tang connection, especially with metal rigging! Metal rigging is very unforgiving when torqued the wrong way, usually resulting in stress cracks.

You would think that these wonderful and relatively inexpensive toggles would be in widespread use, but look around and you will rarely see a boat with such toggles.

Toggles add one more axis of rotation to the stay, which creates a universal joint, allowing the stay and mast to move around freely without causing undue strain on the metals. As the boat sails, the mast moves around and if your stays are connected to the tang directly, they would only have articulation for fore and aft movements. Lateral movements would cause stress on the fittings.

Synthetic stays are much less sensitive to these forces, as they are made of rope, which can move around more easily. Toggles are used more to orient the eyes into a position that avoids chafe, especially on the lowers where the eyes might rub on each other in the tang. Rotating them 90° separates them and prevents any unnecessary chafe.

While toggles on metal rigging should be considered mandatory, with synthetic rigging, they are considered a convenience item and used to help orient the eye splices as they connect to their mast end fittings.

DEADEYE

Deadeyes serve one purpose: to connect the synthetic stay to the chainplate. Chainplates have a small hole in them designed to connect the rigging via a clevis pin attachment. Normally, the clevis pin is connected to a turnbuckle, but with deadeyes, the clevis pin connects a toggle to the chainplate.

Dyneema deadeyes may look fancy with their loops and fittings, but they are actually just a Dyneema grommet with two thimbles in them. The central tie is only there to hold the thimbles in place.

Making a grommet is a tedious task, and making one out of Dyneema proves to be all the more complex. Dyneema is classified as 12 Strand Class II rope, and relies on a long bury to securely hold the splice, but deadeyes need to be short, so we must rely on the lock to hold the load since we won't have the buried length.

A Mobious Brummel works by passing the ropes through each other, causing them to lock against one another when pulled. The tail is then buried, further locking the splice in place. For the junction to open up, the 12 woven strands need to unravel and separate in order to pull apart. The pressure from the woven tube crushing down on the buried tail will not allow the strands to unravel and will keep the splice secure. Locking stitches will add extra insurance to make sure that nothing slips and everything holds.

Mobious Brummel splices are easy to do, simply pass the two free ends through each other and bury the tail. When making a grommet, this is not possible. There is no way to pass the other line through as it is trapped on the other side of the grommet. To get around this, you simply deconstruct and reconstruct the line as you make the splice.

BUILDING IT

The standard way to make a deadeye is to take a 24-inch length of SK78 in the same diameter as your stays and then fold it over in half. The top will remain the midpoint for the process, but the bottom where the legs are falling will be where you will be doing all of your work. Halfway down the legs, you will pass one tail through the other with the aid of a fid. This will be your first cross in the splice. Making a normal mobious brummel eye splice would now cause you to pass the second leg through the first leg to make the second cross, but you are not making an eye, you are making a grommet. You need to now flip the legs around so that the tips are now up near the top midpoint.

Securing the first crossover with a pin is helpful to keep the strands from shifting as you begin making the second crossover. Since your second leg is trapped inside the loop, but needs to end outside of the loop, the only logical answer is to unbraid the 12 strands, pass them around the loop on both sides, and then rebraid them on the other side. Passing six strands on each side will create the same effect as if you magically made the splice happen. If you are a braiding wizard, you can maneuver each of the 12 strands into the perfect cylindrical rope that it once was, but for us mere mortals, there is an easier approach. The six strands on each side can be braided into four groups of three strands. These braids are easy enough to do and make it less strands to keep track of. The four braided strands are then plaited to make a single new rope on the other side of the splice. When you pull on the loop at this point, the lock should hold and keep the splice from pulling apart.

Now the fun part: burying these two tails. The midpoint at the top becomes really important as you want to do everything evenly. Inserting a fid into the grommet will reveal a massive shortcoming to the method, fids are straight and grommets are round. You want to have the tails wrap three-quarters of the way around the grommet, but you simply can't make the fid arc like that. As you open the weave to push the fid through, the line shortens as well. Marking the midpoint is important before you start pushing the fid in as the midpoint will seemingly shift around as you work.

Push the fid in at the base of the tail and then pop it out at the midpoint. Doing this on both sides will make your grommet look like a lollipop with a handle. Now push the fid in at the midpoint and pop it out halfway around again, being careful to run it next to the other bury that is already in there. Follow the fid with the tail and you should now have something that looks like a donut that wants to give you a hug. Pull the tails out as far as you can on both sides and then taper the tails as best you can before you milk the tails back into the grommet.

The goal is that the tails lay inside the grommet and taper next to each other as they cross the midline. This would give a uniform thickness to the grommet, but if it doesn't come out as perfectly, the changes are at least gradual to reduce stress.

Adding some thimbles and tying the middle will give you the finishing touches to turn this really strong rope loop into a deadeye that can connect your stay to the chainplate and allow you to tension your rigging on a beer can budget.

A turnbuckle works by creating an inclined plane by which the end of the stay is drawn down. This incline plane is merely wrapped around a rod known as the bottle screw, which is the threaded part of a turnbuckle system.

The mechanical advantage of an inclined plane is equal to the length of an inclined plane divided by its vertical height. If you have a ½-inch screw with 16 threads per inch, you are looking at a circumference of 1.57 inches per thread and 16 of these per inch of vertical travel. This means that if you stretched out the thread of a turnbuckle's bottle screw, you would have 25.12 inches in length for every inch of height gained. This means that the turnbuckle gives you about a 25:1 mechanical advantage when you are tightening your rigging. This can be further increased by using a wrench on the turnbuckle body.

If the wrench is 10 inches long, that 25:1 mechanical advantage just increases by 10-fold and becomes 250:1, which will certainly give you plenty of pull to get the tension required in your rigging. You might be wondering how on earth could you ever get that kind of leverage without a turnbuckle; the secret is in the multiplier effect of a deadeye pulley system.

In a 2:1 pulley system, the load is divided over two standing lengths that are taking up the work. The weight gets cut in half, but the amount of rope that needs to be pulled is doubled. Adding one extra standing length in the pulley system suddenly increases the mechanical advantage to 3:1 where the weight is reduced to one-third of its original weight and the length that needs to be pulled is tripled. Basically, you are trading weight for length of rope that needs to be pulled, and by simply adding another standing part you are able to increase the mechanical advantage monumentally.

The multiplication of pulleys and their mechanical advantage can be taken one further by stacking the effects and thereby multiplying the multiplier.

For a standard lower shroud, I recommend just four passes of the lashing between the deadeye and the end of the stay. This will create an 8:1 pulley system. The lashing line is then connected to a tensioning line, which is another 2:1 creating a 16:1 total system. The tensioning line is further connected to a winch that increases the mechanical advantage by the number stamped on the winch. If you have a Number 45 winch, you will be multiplying 16:1 by 45, which will result in a staggering 720:1. This is far greater than the simple 25:1 of

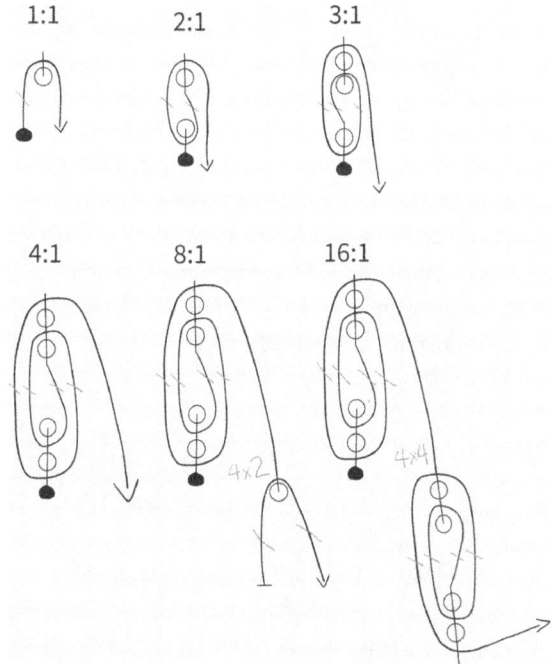

a turnbuckle or 250:1 when using a wrench! Naturally, a turnbuckle only has the frictional losses of the bronze turnbuckle body on the stainless steel threads of the bottle screw, so the full mechanical advantage is not realized, but it is very close to this amount of a force multiplier. The lashing system is much more involved and as a result the amount of loss due to friction will be much higher, but you will still be able to achieve these staggering numbers with a very simple pulley system and that will give you the tension needed to tighten your rigging to the required amount.

For cap shrouds and headstays, I recommend increasing the pulley system to seven passes of the lashing between the deadeye and the end of the stay, which will give you a 14:1 pulley system, once again acted on by a 2:1 with the tensioning line, resulting in a 28:1 pulley system, which is fed into a winch. If you have the same Number 45 winch, then your system would tower over the lower shrouds system with a staggering 1,260:1 mechanical advantage. Naturally, there will be losses to friction so every pound your arm puts into the winch handle doesn't become 1,260 pounds of pulling force, but it will certainly be more than enough to tension the heavy shrouds to the specified tensions. As you can imagine, the bigger the boat, the bigger the rigging and therefore the greater the tension on the rigging needs to be. At the same time, the bigger the boat, the bigger the winches, allowing you to tension the rigging on any sized boat that has winches using a deadeye system.

At this point, you know how to splice synthetic rigging, make a deadeye, and set up a lashing system to tension the stay, but how long does each piece need to be? Knowing that your cap shroud needs to be 50 feet from mast tang to chainplate doesn't do you much good when your cap shroud consists of the stay, the lashing, and the deadeye, not to mention all the toggles that go into it as well. What cut length of rope do you need to construct the stay?

Remember that you need to measure and mark everything before you begin working the line. The moment you open the weave to perform a splice, you will lose your ability to estimate the accurate length of the finished stay. The length of the stay needs to take into account the amount of line that will be buried back into the line, as well as the amount of line that will be consumed in passing around the eye, and the length of the stay that will be composed of the lashing and deadeye.

The simple method to calculate this is:

Length of the stay + Perimeter of thimble + Amount buried − Length of deadeye − Length of lashing

The length of the stay is the distance from the mast tang down to the chain plate.
The easiest way to calculate the perimeter of the thimble is by wrapping the line around the thimble and measuring the length involved.

The amount buried is 72 times the diameter of the line in millimeters. You will also want to taper the end of the tail that gets buried, so it is wise to give yourself double the minimum amount of bury, having the minimum length with no taper, and the next minimum length tapered.

The finished length of the deadeye tends to be around 1 foot.

The length of the lashing is up to you, but I recommend planning it to be around 3 feet. This distance gives you plenty of leeway in case your stay comes out a bit long, and it places the lashing in harm's way of the lazy sheet. This means chafe will occur on a disposable lashing instead of an expensive stay.

Once you have all of these numbers calculated, it is time to mark it on the line before you cut it off the spool. When you mark your points, you can then make sure that everything is correct before you cut the line. If you cut the line too short, you will not be able to make the stay the correct length. If you cut the stay too long, you can always trim it shorter. This is why I like to cut a few extra inches of leeway on each end just in case I find that I need a bit more length.

Let's work through the math for a 9mm line:

The first point will be the tapered length. To get a nice gradual taper, you want to taper the tail over the length of the minimum bury, which would be 72 × 9mm = 648mm or 25.5 inches

For standing rigging, you want to bury double the minimum length, tapering one minimum length and having the other minimum length buried at full thickness. This gives you overkill on the buried portion and makes your stay incredibly strong. As calculated previously, 648mm or 25.5 inches for this section of the rope. This makes the total length of the buried tail 1.3m or 4 feet, 3 inches long.

At this point, you have the taper and the bury marked on the line. Now you will need to measure and mark the thimble section. The most accurate method is to place the throat of the thimble at the tail mark and work the line around the thimble, then mark it again. You want to make the eye that goes around the thimble loose, that way you can easily replace the thimble if it fails as the years march by. The looseness of the eye can be closed by a small seizing knot to hold everything in place during installation.

All this length does not contribute to the actual length of the actual stay. Now that the tail portion is calculated, measured, and marked, it is time to measure out the length for the actual stay.

The length of the stay is the span from mast tang to chain plate minus the four feet of the lashing and deadeye. In this example, that distance is 50 feet. If you are going to have 3 feet of lashing and 1 foot in the deadeye, this would mean that $50 - 3 - 1 = 46$ feet of length.

Your length of line would be as follows:

- 2 feet for taper
- 2 feet for bury
- 8 inches for thimble
- 46 feet for stay
- 8 inches for thimble
- 2 feet for bury
- 2 feet for taper

As you can see, the cut length is 55 feet, 4 inches long.

When you finish fabricating the stay, you will notice that it will be much shorter than 46 feet long that you were trying to achieve. This shortness is due to the constructional shrinkage, which at this time can be thought of as potential constructional stretch.

The more weave you opened to perform the splice, the shorter the stay will appear to be. Upon initial tensioning, this stretch will be removed and it will nearly approach the ideal 46 feet of our example. After a bit of creep during phase one of the Dyneema lifecycle, you will find that it will reach closer and closer to the ideal 46 feet desired.

This is very important when working with headstays, where you want the eye splice to sit as low to the deck as possible, allowing you to have your first hank attach to the stay as soon as possible. Proper measuring can safely place this eye at the exact length to engage a turnbuckle if you choose to use one.

If you have to err, it is best to err on the short side. If your stay is too short, the stay will have a longer lashing and will still achieve the same amount of proper tension. If your stay is too long, you will end up two blocked and you will need to redo your splice and tension it all again.

Headstays and shrouds that will come in contact with sheets or sails will get chafed. This chafe will prematurely destroy the stay and is also easily protected against by simply installing a chafe sleeve over the stay when it was being spliced together. The length of the chafe sleeve is equal to the length of the body of the stay between the eye splices. Once the splices are all finished, the chafe sleeve is merely sewn into place just under the throat of the eye splice. For areas that will see more severe chafe, service can be applied. One such spot is the spreader tips, not only for the chafe protection but for the added friction to make it easier to tie up and seize the shroud to the spreader tip. Lastly, when installing the stays, be sure to look up the stay and untwist the rope. 12 Braid rope is torque neutral, which means that it has no twist. When you look up the stay, make sure that the braids run up in a straight line as opposed to twisting their way up. Minor twisting doesn't have much of a structural concern to it, but it just makes it look like the job was done rather carelessly over the next several decades as someone looks up the mast to inspect the rigging.

You have spent the first half of this book learning how everything works and how it should all be set up. You will now begin learning about how things fail and what we can learn from these failures to prevent them from happening again, as well as how to make sure your work won't suffer these same problems.

PART II
MECHANICS OF RIGGING

CHAPTER ELEVEN
WHEN THINGS GO WRONG

THERE ARE TWO METHODS FOR DEALING WITH RIG FAILURES. THE FIRST IS TO notice the fault and repair the rig before anything happens; the second is to wait until it breaks and then use what you have left to jury rig something that will get you home. The purpose of this book is to educate you so that you can spot major problems when they are in their infancy, rather than letting them develop into unruly teenagers who will wreck your boat as well as your dreams!

When things are about to go wrong, they first present themselves in the quietest and most polite way possible. Breaking clocks don't make a sound, and neither do failing rigs! The trick is to learn what to look for so that you can recognize these subtle clues and avoid the terrible results that come from neglecting your rigging.

The most common presentations for future failure are corrosion, cracks, deformation, chafe, ultraviolet damage, and a loss of tension. While most of these changes are very minute, they can all be seen by simply paying close attention. Loss of tension is somewhat a visually subtle occurrence, but a very obvious issue when you simply give the rigging a wiggle and a tug.

CORROSION

Corrosion is an electrochemical oxidation of a metal where it reacts with an oxidant, most commonly oxygen, but also hydrogen or hydroxide. The most commonly known form of corrosion is called rust, which is actually a special kind of corrosion where iron forms iron oxide. Iron is the main ingredient in steel, which makes up so much of the world we live in, so rust is the most commonly experienced type of corrosion in a layperson's daily life. The use of materials that contain iron is strictly limited on a boat as it readily reacts with the marine environment to corrode away.

At its heart, corrosion is an electrochemical reaction where the metal acts as an anode and gives up its electron to reduce an oxygen in the presence of H^+. You need three parts to make a battery, the anode, the cathode, and the electrolyte. The anode gives up the electron, the cathode accepts the electron, and the entire process is facilitated by the electrolyte solution. Over time, the anode gets consumed and dissolves away, which is why batteries don't last forever. On a boat, there is plenty of H^+ present in the seawater, and that fresh breeze is packed full of oxygen! In other words, your boat is just floating inside of a battery and the metals on your boat are the anode, slowly being eaten away as they are converted from their pure metallic form to an oxidized form that will no longer have the necessary properties to support your rig.

RUST

Rust is the process of iron (Fe) turning into ferrous oxide ($Fe^{+2}O^{-2}$) or ferric oxide ($Fe_2^{+3}O_3^{-2}$), which is facilitated by the presence of oxygen in a moist environment. Salt spray helps add Na^+Cl^- to the mix to speed up the process of forming oxides. This only affects iron-containing metals and will not affect bronze (a copper alloy) or aluminum parts of the boat. Stainless steel is an alloy of iron with carbon, but to make it "stain-less" they add in chromium, which forms an oxide layer on the surface of the metal that isolates the underlying iron from oxygen and therefore prevents rust from occurring as readily.

GALVANIC CORROSION

The process of making a battery as outlined previously talks about needing an anode and a cathode in an electrolyte solution. In the previous situation we really only had an anode as the oxygen was filling in for the cathode in that scenario. With galvanic corrosion, you have the full-on battery situation! When dissimilar metals are placed in contact with each other and then immersed in an electrolyte, such as salt water, you will have a flow of electrons from the anode to the cathode. Over time, the anode will get eaten away and become weakened while the cathode remains just fine.

This is precisely why underwater metals on boats have a sacrificial anode attached to them, that way something easily replaceable can get destroyed rather than an expensive and important part of your boat.

In general, the least noble metal will get consumed as it will act as the anode and the most noble metal will survive as it will act as the cathode. In the simplest form, if the metals are more than two apart on the table below, they will cause problems and another less noble metal needs to be added to the mixture to act as the sacrificial anode.

Galvanic Series
Least Noble

Magnesium
Zinc
Beryllium
Aluminum Alloys
Cadmium
Mild Steel & Cast Iron
Low Alloy Steel
Austenitic Cast Iron
Aluminum Bronze
Naval Brass, Yellow Brass & Red Brass
Tin
Copper
Admiralty Brass, Aluminum Brass
Manganese Bronze
Silicon Bronze
Stainless Steel: Grades 410 & 416
Nickel Silver
90/10 Copper Nickel
80/20 Copper Nickel
Stainless Steel: Grade 430
Lead
70/30 Copper Nickel
Nickel Aluminum Bronze
Nickel Chromium Alloy 600
Nickel 200
Silver
Stainless Steel: Grades 302, 304, 321 & 347
Nickel Copper Alloys 400 & K500
Stainless Steel: Grades 316 & 317
Alloy 20 Stainless Steel
Nickel Iron Chromium Alloy 825
Titanium
Gold & Platinum
Graphite

Most Noble

The stainless steel shaft of the boat attaches to the bronze propeller, with 13 other metals between them on the galvanic series demonstrated above. As a result, the bronze of the propeller will get consumed away as it will act as the anode in the battery. To save the bronze, an even less noble metal is employed as the sacrificial anode, commonly referred to as "zincs" but they can be magnesium, zinc, or aluminum. They are affixed close to the bronze so that it will get eaten away before the bronze does. The important part here is you need to watch the condition of your zincs as when they get eaten away, they stop working and the bronze begins to get consumed!

Ironically, steel is very far away from aluminum on the galvanic series, which is why steel sailors always laugh a little on the inside when they see an aluminum sailboat in their anchorage, as they are looking at their new anode in the water.

Looking at the chart, you can also see how people get into trouble when they attach a few different types of bronze alloys together. They have different mobilities and therefore will react and damage each other over time.

Lastly, you can also see the extreme difference in nobility between 316 stainless steel and aluminum, which is why stainless-steel fasteners that are used to attach components to a mast need to be isolated from one another or the aluminum of the mast will quickly begin to bubble the paint around the steel screw as the aluminum spar turns into a white powder known as aluminum oxide.

The best way to prevent this kind of reaction is to never mix different metals. Always attempt to use fasteners that are made out of the same grade of metal as the item you are attaching; also always attach the same metal onto more of the same metal. Being that this is not always possible, the second-best thing is to use something as an electrical insulator between the two metals, thus preventing the formation of an electric charge and, as a result, preventing the deterioration of the least noble metal in the mixture.

Metals can be isolated with fancy synthetic materials such as teflon, which can be used in a sheet form or as a gel form, think really wide teflon tape or TefGel; or they can be isolated by much more natural substances such as Lanocote, which is nothing more than the lanolin from sheep's wool. Personally, when I'm doing jobs that are closer to the deck where inspections are easier to carry out, I prefer to use Lanocote as it is non-toxic and if you get it on your hands, your hands just feel well moisturized. If it washes out over time, it can easily be reapplied by removing the fastener, dipping the threads in the jar, and re-attaching the fastener. On more permanent installations, I use TefGel as it is synthetic and will last a lot longer. Lanocote is significantly less expensive than TefGel, and works just as effectively, so it is always a good alternative for small projects near deck level on the boat.

CREVICE CORROSION

If corrosion were a belly button, galvanic corrosion and rust would be "outies" while crevice corrosion would be more of an "innie." Crevice corrosion is the degradation of the metal deep beneath the surface, slowly eating away at the metal from the inside until it becomes so weak that it simply breaks!

Stainless steel, titanium, and some bronze alloys rely on a superficial oxide layer to protect the underlying metal from oxidative corrosion. As long as oxygen is present, metals in the alloy will react with the oxygen to form a protective oxide layer. Stainless steel accomplishes this with chromium, which forms a chromium oxide layer, while aluminum bronze accomplishes this same task with the aluminum forming an aluminum oxide layer. Regardless of metal that is employed for the task, the process is the same.

The oxide layer acts as a shield keeping water, oxygen, and any other nasties away from the metal. This process works just like paint on steel to keep the metal isolated and prevent rusting, but unlike paint, if the surface gets compromised, it will instantly react with more oxygen to heal the surface and reinstate the shield for the metal below. Oxygen is a very important part of this process, which is why if the metal exists in an anaerobic environment or simply one with low oxygen levels, the protective layer is unable to form and the metal becomes unprotected. This kind of unfortunate circumstance is readily present on a boat. Internal chainplates that are placed inside lockers or behind the boat's furniture will live in a realm where there is less air turnover. As a result, the oxygen in the local air can get used up until it is no longer enough to produce the protective oxide layer that is needed for the metal to live up to its touted fame.

The other offending situation is intentionally caused by us as the internal chainplate passes through the deck. The sealant that is placed around the chainplate to keep water out of the boat also keeps all the oxygen away as well. When the sealant is new, it sticks to the outside of the chainplate, which keeps everything away from that metal. As the sealant ages and degrades, it will pull back slightly from the surface of the chainplate. The obvious result of this is the dreaded "leaking chainplate," which will drip water into the cabin. The more nefarious aspect is the chainplate as it passes through the deck that is in an oxygen poor environment and being exposed to saltwater while not having the protection from the oxide layer.

To further the problem, seawater acts as an excellent electrolyte as seawater is water (H_2O) and salt (Na^+Cl^-) that will dissociate in the water to form a solution of hydrochloric acid (HCl), hydroxide (HO^-), and hydrogen ions (H^+). From chemistry class, you might recall that the pH of a substance is lowered by the presence of hydrogen ions. In other words, seawater is nothing more than acid water on your metals!

The only thing worse for the metals on your boat than acid is hot acid, and that is exactly what happens when the sun beats down on your deck.

Seawater that has creeped into the little nooks and crannies between your degrading sealant and chainplate will then get heated up well over 160°F (71°C), which is hot enough to speed up the chemical process and cause lots of problems.

The issue goes something like this: The oxide layer gets compromised, but only in a little spot. Hot salty water gets in there and begins to attack the metal in the unshielded region, which creates a tiny pit and this will serve as the origin for the ending of the tale of this piece of metal. This pitting will then continue inward on the metal as the pitting process continues over and over inside the metal at an alarming rate. Pitting is visible on the surface, but only the most superficial pit, the rest of the lattice network inside the metal is obscured from view. Crevice corrosion can extend beyond the pitted area and travel deep through the metal, effectively cleaving the metal. The remaining metal is forced to support the load as this invisible crack compromises the strength of the chainplate inside the deck, hidden away behind the sealant.

Let us review: The chainplate is mounted inside the boat to a bulkhead or some other structure, it then passes through the deck, which would let water into the boat so it is sealed. The internal portion of the chainplate isn't very pretty to look at so it is hidden away behind the beautiful interior of your sailboat causing it to live in an oxygen-poor environment, which will lead to a breakdown of the protective oxide layer and the potential for crevice corrosion to develop. The sealant begins to degrade and pulls away from the chainplate slightly exposing the surface of the metal, which has a weakened oxide layer due to living inside a sealed area with minimal airflow, allowing the seawater to get in that tiny gap next to the chainplate where the sun heats the metal and raises the local environmental temperature to very hot temperatures. The hot acid begins to attack the surface of the metal and causes small pits to form on the surface of the metal. The process continues for a while until the sealant finally breaks down enough that water is allowed to seep into the cabin area of the boat. Now saltwater is slowly dripping onto the chainplate inside the boat, onto the metal that has had a weakened surface coating due to living in the oxygen-poor environment that exists behind your cabinetry. The salty water will dry and leave behind sodium chloride crystals that will attract moisture from the air and dissolve into free sodium and free chloride ions. The chloride ions are the ones that wreak havoc on the metal, and they will continue to do so until you finally notice that your chainplates are leaking. You reseal your chainplates with new sealant, but the salty solution continues to live on the surface of the metal, now sealed in for good measure. Humidity in the air keeps the ions dissolved as they create incredibly acidic solutions whose action is expedited by the heat of the sun beating down on the exposed metal of the chainplate. In time, the pitting turns into deeper crevice corrosion and the chainplate is done.

The worst part is the only way to properly inspect the chainplates involves pulling them yearly, and replacing them if they show any signs of degradation as that is the only hint to a deeper lurking critical failure. If you have never removed a chainplate, then you might not understand why this is such an arduous request because it is a lot of work to remove a chainplate! The thought of having to do this yearly might turn some people off of sailing all together. As a result, the chainplates are normally ignored for a decade and hopefully replaced along with the standing rigging.

As you can imagine, having external chainplates helps to mitigate a lot of these issues. The chainplate is, first and foremost, much more accessible to visual inspection; second, it is exposed to all the oxygen, which helps keep its oxide layer working at peak performance. The only issue that can arise is crevice corrosion from the side that faces the hull if the sealant were to break down, which simply means that every few years it would be prudent to remove the chainplates and rebed them. This would be a preventative precaution as you are removing and replacing the bedding compound or sealant before it fails, that way the inner side of the chainplate remains hidden from salty spray and moisture.

My favorite way to look for crevice corrosion is to shine a very bright light at the surface of the metal and have the light reflect into my eye. As you can imagine, this is not a pleasant method, but it will make the entire surface of the chainplate appear as a bright white field. If there is a small crack in the surface, the crack will show up as a black crack in the bright whiteness of the chainplate. Moving the light around as you stare into its reflected brightness will help make these minute fractures appear more easily before you.

There are a multitude of other testing methods that can be employed but none of them are available to the average sailor. X-rays, sonic testing, and submerging the metal in chemicals to exhibit the presence of the corrosion are some of the professional methods that can be employed to diagnose crevice corrosion, but these simply are not available to most sailors, which is why we need to rely on other, more primitive methods such as "giving it a good look."

CRACKS

Cracks in rigging appear for a few reasons, but all of them will indicate to you that you need to replace that part of the rigging immediately. There are two orientations of cracks that you will commonly find on your boat: vertical and horizontal.

Vertical cracks will tend to appear on swaged fittings at the end of a stay, and they mean that the cable has corroded in the swage. As the metal corrodes, it occupies more space but is trapped within the confines of the swage terminal with nowhere for the developing oxide to go. The pressure inside the swaged fitting increases until the pressure exceeds the capacity of the metal and it cracks. A small vertical crack line running along the side of the swage is all you will see, and if you see this, then you know the swage is cracked and no longer holding the rigging wire in place. With enough force, the wire will slip out and your mast will lose that stay. As a general rule, if you find one cracked swage, you can assume that the others will be cracking soon and it is best to replace all your standing rigging immediately.

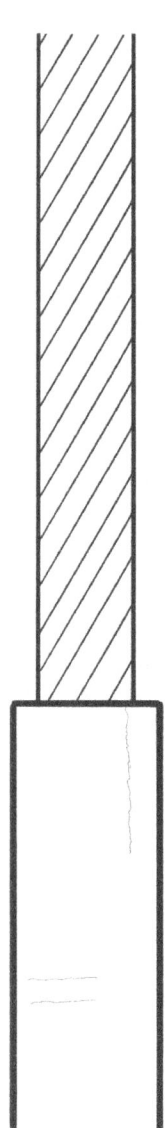

Horizontal cracks run side to side instead of up and down. These form for a few reasons, but generally are the result of overloading. Stainless steel is an incredibly strong and hard metal, as a result, it will yield just a moment before it breaks. The yield strength for stainless steel is around 36,000 pounds per square inch (psi) while the ultimate tensile strength is around 82,000 psi. The Brinell hardness for 316 stainless steel is 210, which means that this metal will resist deformation rather well, which also means that it won't show signs of deformation as easily. Most loads on a sailboat are static loads, picture the boat heeled over while sailing on a close reach where the rigging is loaded and everything simply remains that way for hours or days. Shock loads, however, picture an accidental or crash jibe, are much less predictable in both occurrence and force. When the boom flies over and crashes onto the other side, the shock load on the rigging is nearly impossible to anticipate and size the metals accordingly. For this reason, safety factors are added to the calculations so that these unforeseen forces are hopefully within the realm of what the metals can withstand. As a result, sometimes they will exceed the yield strength of the metal and cause it to deform, but most of the time they will approach the tensile strength of the metal and cause it to break. The cracks will form perpendicular to the direction of the force, which is why these cracks appear as horizontal fracture lines.

These cracks are a subtle reminder of a time when the rigging was literally pushed to its breaking point, and while the parts are still presently holding your mast up, you should thank your boat's guardian angel for only giving you cracked fittings and letting you keep your mast. Luck only goes so far, which is why you need to replace these components immediately with new ones and also evaluate the scenario that led you to perform a crash jibe or other uncontrolled maneuver so that a plan of action to avoid repeating this mistake again in the future can be made.

WHEN THINGS GO WRONG

While swaged terminals will form horizontal fracture lines from being overloaded, the chainplates are oversized significantly and therefore won't show horizontal crack lines for the same reason. If you overload the rigging, the excessive force will crack the swage or break the wire, but the chainplates are so strong that they won't suffer much from this event. As the metal degrades due to pitting and crevice corrosion, the metal will weaken to the point that these high loads can now cause cracks to form. The spot to look for these cracks is at the top chainplate bolt hole or at the top of the chainplate where the stay attaches via a clevis pin. The sides of the chainplate next to the hole have a lot of load to carry and less metal than the area above and below the bolt hole. If the metal was weakened by the presence of crevice corrosion, then a visible crack line would develop. Once again, this is your very subtle warning sign that the end has come for the service life of your chainplate and it needs to be replaced. As a general rule of thumb, if one chainplate is exhibiting the signs of failure, you can safely assume that the other chainplates are failing as well but just not showing any outward signs yet, which is why it is best to go ahead and replace them all at once. Replacing all of them at the same time also makes it easier to manage the maintenance schedule of the boat, as chainplates and standing rigging are to be replaced every 10 years, 10,000 nautical miles, or when signs denote that replacement is needed earlier.

Wire rope has been the standard material to connect the long portion between the chainplates and the mast fittings for over a hundred years. The type of wire has changed only a few times as metallurgy improves. The three types of wire rigging you will come across in current times are 7×7, 7×19, and 1×19. The naming system for wire rope is very simple as it simply tells you how many strands and wires the rope is composed of.

The more wires in total, the more flexible the wire rope will be and the more forgiving it will be to take a turn or bend. More flexible wire rope can be used in pulley systems involved with steering and swing keel mechanisms. Less flexible wire rope is relegated to only serving as standing rigging as it will, for the most part, run in a straight line.

The naming system tells you a lot about the wire rope, as the pure algebraic conclusion will tell you the total number of wires in the rope.

$$1 \times 19 = 19$$
$$7 \times 7 = 49$$
$$7 \times 19 = 133$$

The first number indicates the number of wires that each strand is made of, and the second number indicates the number of strands wrapped together to make the rope. The name is spoken *First Number by Second Number*. It is rather rare to find a boat with 7×7 or 7×19 wire in use for the standing rigging anymore, but if you do, pay special attention to the top and bottom of the stay as it will most likely be an eye-splice. These wire ropes take a bend well and can be spliced around a thimble to terminate as an eye. Splicing wire rope is an incredibly difficult task, not because it is a difficult concept but rather because the wires are stiff and getting them to go where you want them to is a tricky task. There are also a lot more strands to manage and if you get it wrong, undoing the splice to correct the mistake is almost as difficult as getting the wires into those places the first time! For this reason, you will rarely see a wire splice, unless you are at a museum or meet a sailor with an affinity for the old ways and did it themselves.

7×7 and 7×19 wires are still used for lifelines, but even this purpose is being phased out in favor of 1×19 or even synthetic lifelines. When 7×7 and 7×19 wire was used for standing rigging, sailors would rub a cotton ball along the length of a stay to check for broken wires, known as "meat hooks." The individual wires are so small and thin that it is hard to see them individually, and if you ran your hand along the wire to check for any broken strands, you would surely have pieces of flesh ripped out of your palm! The cotton ball is a handy trick as it identifies the broken wires without putting you at risk of bodily harm. 7×7 and 7×19 wire was traditionally galvanized, and to protect the metal from the harsh marine environment, it would be kept wormed, parceled, and served and coated with slurry. Galvanized wire that is entombed inside this oily realm will last for decades and still look brand-new in there, as long as it is maintained on a close and regular schedule.

COURTESY OF JACOB BLACK

If the wire is not properly maintained, then you need to worry about corrosion and meat hooks, as these are the signs of failure. As you might have guessed, on wire rope rigging with many small strands, a horizontal fracture shows up as a meat hook. On 1×19, where each wire is much thicker, it won't curl back on itself to make the hook part, but it can still break.

A broken wire on 1×19 will typically occur near the swage fitting, as this portion is work hardened as the metal leaves the grip of the swage and any bending action falls on this small span of wire. When the wire breaks, it will present itself in two ways: either a small crack in the wire right at the swage or it will unravel a little and lift itself up, appearing as a "raised wire." A single broken wire represents about a 5% loss of strength, which will overload the remaining wires and lead to catastrophic failure if this warning sign is not heeded. Both of these breaks are presentations of horizontal fractures in the wire.

DEFORMATION

When metal is subjected to extreme loads, it will bend before it breaks. How much it will bend and how far before it breaks depend on the properties of the metal. Alloys with high hardness will resist deformation and will usually hold their shape up until the very end, giving little warning that they are being overloaded until the catastrophic end. Softer metals will deform well before they break, which offers more of a warning during inspections.

During a rig inspection, you need to look closely at all the individual components as you are looking for any sign of bending, known as deformation. Clevis pins, swages, toggles, and everything employed in the rigging should look as it did the day it was installed. If a bail looks a little oblong, it has probably been pushed to its breaking point and needs to be replaced. If the replacement suffers the same fate, then that component might be undersized or it is being misused and therefore overloaded.

Bronze and stainless steel are the two most common metals you will find in use for standing rigging. Bronze is a softer metal than stainless steel, which is why bronze fittings will bend earlier than an equally sized stainless fitting. Since they will both break at the same point, both metals can be used interchangeably in the same size. The real difference between the two metals is that stainless steel will look shiny longer, stainless steel will suffer from crevice corrosion, and stainless steel will hold its shape until the very end. Inspecting stainless steel is a challenge as your warning signs are slight cracks on the surface, while inspecting bronze is easy as you are looking for deformation from the original shape.

Deformation is possible to find in stainless steel and it will usually be accompanied by cracks as the metal was pushed to its breaking point. In the table below, we will go over each part of the rigging and highlight what to look for in terms of deformation and cracks depending on which metal you are investigating.

As mentioned in chapter 8, the clevis pin acts as a fuse for the entire rig, and should the load supersede what it can safely manage, the clevis pin should shear and cause the rig to break free, and also come crashing down. Naturally, we don't want the rigging to have any weak points that would give way and cause the mast to collapse, but without a designated "fuse" the next weakest point will give way and cause the rigging to fail. If the rigging never failed, the rest of the boat would be in danger as it could be pulled under should the sails be filled with way too much wind or, worse, water, which would hold the boat on its side and not allow the boat to right itself. Eventually, the hull would fill with water and sink, which would be much worse than simply losing the mast! Having a predetermined and calculated point of failure is crucial to building a reliable and seaworthy system, which is why clevis pin size should be respected and maintained. It is also why upsizing the wire size "to make the rigging stronger" is rather pointless as the clevis pin size will remain the same.

Component	Warning Signs in Stainless Steel	Warning Signs in Silicon Bronze
Chainplate	Horizontal cracks at the top bolt hole. Deformation and cracks at the clevis pin hole.	Deformation at the clevis pin hole. The round hole will look ovate as the top was pulled upward. The top of the chainplate might have a bit of a bulge to it compared to the other chainplates.
Clevis Pin	Bent shaft and cracks at the height of the bend.	Bent shaft.
Cotter Pin	Broken legs from work hardening.	Broken legs from work hardening.
Toggle	Deformed clevis pin holes, ovate hole instead of perfectly round hole. Cracks around the clevis pin hole. Cracks on the body of the toggle. Cracks along any bend of the toggle.	Deformed clevis pin holes, ovate hole instead of perfectly round hole. Stretched toggle body, notable by comparing length to that of a new one. Logo or writing on the body may appear to be stretched oddly.
Turnbuckle Body	Cracks forming in the body. *Most turnbuckles are actually chrome-plated bronze.	Stretching of the turnbuckle body.
Swage Fittings	Horizontal or vertical cracks.	Not normally used for swage fittings.
Wire	Broken wires near swage fitting.	Not normally used for wire.
T-Ball	Cracks in the area of the neck.	Not normally used for T-ball fittings.
Eye Terminator	Cracks around eye fitting.	Not normally used for eye terminators.
Mast Tang	Cracks around the clevis pin hole.	Deformation around clevis pin hole. Stretching of the mast tang.
Padeyes and Bails	Cracks along the hoop. Rarely a bend in the hoop.	Bent hoop with a sharp point at height of the bail. One giveaway is if a component always rests in the exact spot, try to move it out of there and you will see a little depression from where it has deformed the metal.
Gooseneck	Cracks in the fittings and connections.	Deformation in the fittings and connections.
Stem Fittings	Cracks or deformation around the clevis pin holes and base.	Deformation around the clevis pin holes and base.

As rigging evolves over the years to come, I doubt that the clevis pin will disappear from the system as it is such a valuable tool that is both easy to inspect and replace, giving warning to the forces the rigging has been subjected to, and making the final connecting of the rigging quick and easy.

Stainless steel clevis pins are the norm, while bronze clevis pins are far harder to come by, but can be found on the internet. Bronze clevis pins are also more expensive, which might explain why they are not as popular on the shelves of chandleries or on sailboats, but they will give you an earlier warning that the loads on "that stay" have gotten pretty high up there and it would behoove you to think about when that might have occurred, taking steps to prevent that scenario from repeating itself as you replace it with a new one.

CHAFE

When rope rubs against anything, the friction can cause the fibers on its surface to break. When enough fibers break apart, they will form fuzzy warning signs that will alert you to a potential point of failure. Chafe can occur from two ropes rubbing against each other, commonly seen when dock lines cross each other, or when a line doesn't have an appropriate fairlead that is causing it to rub hard on a sharp corner. If the rope is forced to bend too abruptly or is rubbing hard against a surface, be it smooth or sharp, the line will eventually begin to break down and chafe will become apparent at this point.

On running rigging, chafe is usually seen where sheets run through blocks that are causing the line to bend at too steep of an angle, as well as where the sheet rubs against a part of the boat. Dock lines are another type of line where chafe is frequently seen as the lines sometimes rub on toerails or other lines as they traverse the path from deck cleat to dock cleat. One last and less obvious place to find chafe is on a halyard where the line passes over the sheave. The halyard will rest on that one spot on the sheave for years of service, as well as the positions that match up with the reefing points, and these spots can begin to chafe. The last and most detrimental position to find chafe is in the anchor rode, as this one single line is responsible for holding your boat in place and preventing it from drifting away! If the anchor rode chafes, the line can break and your boat will be adrift. You will often see boats on the banks of channels or up on beaches with the anchor line still coming out of their bow roller to a short frayed end.

Synthetic standing rigging can also suffer from chafe if sheets are set to rub against the stays. This constant rubbing will cause the fibers in the stay to break and form a nice fuzzy layer. As inviting and comfortable as a fuzzy layer might sound, it is a very bad thing to see in your rigging!

Chafe can form externally as well as internally. External chafe is caused by obvious reasons where something was rubbing against the surface of the line while internal chafe is caused by the line rubbing against its own strands. Picture the anchor rode situation where the waves surge the boat backwards and you hear the anchor line creak as it stretches. That creaking sound is the sound of the fibers rubbing against each other as the line straightens out, and that rubbing creates heat, which further degrades the structural integrity of the line. With enough duration, the fibers will lose enough strength to the point that they will begin to break from the inside of the rope. The end result is a chafed line that failed to do its job!

Internal chafe can also be caused by debris such as salt, sand, and dirt that gets into the line and rubs against the fibers as the line is worked. This is why most running rigging has a service life of ten years, because after a decade it has probably sustained enough insult and injury to bring its end around soon. The goal is to replace lines before they break; that way you don't have any mishaps or catastrophic stories to tell at the yacht club bar next weekend.

When inspecting your running rigging, everything should look the way it was when it was first installed; ropes should be clean with vibrant colors and clearly defined edges. As algae and sediment build up on the line, the colors fade away to a green-gray that eventually devolves to black. While the surface color might appear as a cosmetic concern, the real problem is the fact that impurities have penetrated into the line and will cause it to chafe from the inside where it is nearly impossible to inspect. Cleaning the surface of the line with a hose and brush will only push impurities deeper into the line, so while the outside of the line may appear cleaner, the center of the line is still contaminated and filthy. Cleaning a line in a washing machine is strongly discouraged as the motion of the machine will destroy a line in a hurry! Cores will hemorrhage and splices will come apart in a matter of moments inside a clothes washing machine. If your line survives the washing process, you will now be faced with a tangled mass that will be more work than it's worth to untangle. The final reason that washing lines is practically forbidden is because the chemical agents in most soaps and detergents will degrade the line itself, further weakening the fibers and speeding up their eventual demise. The color and appearance of a line should not be looked at with cosmetics in mind, but rather an indicator for the condition of the line as a whole; if it looks really bad, it is time to replace it.

WHEN THINGS GO WRONG

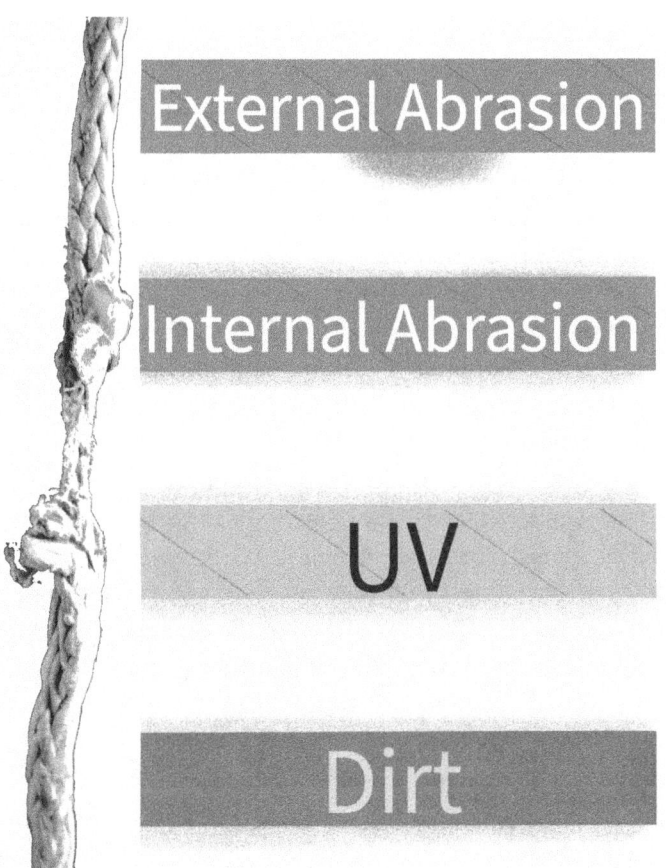

UV DAMAGE

The sun emits powerful amounts of energy that rain down on our planet constantly. A small portion of the energy is in the wavelength range that our eyes can detect and it is called visible light. The rest of the radiation that bombards the surface of our world is not visible to our eyes, but is still coming down with the same power and intensity as the visible light we see. Just beyond the violet wavelengths is the ultraviolet range that we can't see, but we can certainly feel. UV radiation is what breaks down and bleaches anything that is left in the sun for a long time. The reason your child's toy that got left in the yard for a week before it was found turned white and brittle is because it was under the punishing rays of the sun without any protection.

Your sailboat lives its entire life in these conditions with the sun barreling down on the deck, roasting everything in sight to a crisp! The only way anything can survive on this unforgiving deck is to be resistant or protected against UV radiation. Your sails employ a sacrificial layer that acts as a sun cover, shading and protecting your delicate sail cloth from the harmful rays while waiting at its mooring for the next time they will get used. Your rigging, however, has no such sacrificial layer that can shield it and must rely on chemistry to protect the line.

Marine grade ropes have covers that are UV stable to protect the core inside from the sun's rays. This workaround lets double braid ropes be made out of any material desired, even ones that are not UV stable, because the cover will block the sun and shield the core beneath. Single braid lines don't have this luxury and need to be made out of UV stable materials so that they can manage their load and also stand up to the harsh rays all on their own. The nice part about single braid line is when you look at the line during an inspection, you are looking directly at the working part of the rope and know what the status of it is; on a double braid line, you are looking at the cover, which means that you need to rely on hints and clues as to the condition of the hidden core beneath.

Running rigging tends to use a lot of double braid lines as the polyester cover offers great color options to identify each line back in the cockpit, as well as offering excellent hand-feel characteristics when the line is being worked. Lastly, the polyester cover works really well with self-tailing winches, holding well to the drum and not fouling on the tongue of the self tailer. For all of these reasons, most running rigging will be double braid line. Those who are seeking the highest performance will sometimes remove the cover in certain areas of the line in an effort to reduce weight and windage, but they will still keep the cover on in the areas they handle because of these wonderful properties that have made double braid lines so popular.

Synthetic standing rigging on cruising boats is dominated by a single braid fiber known as Ultra-High-Molecular-Weight Polyethylene, or UHMWPE. They are sold under two main brand names: Dyneema and Dux. These fibers have incredible resistance to chafe and UV damage, meaning that they can be used without a cover for their entire service life. The benefits of using UHMPE without a cover is that you can visually inspect the working part with ease. Chafe will show up with a hairy texture over the line, and from the outward appearance of the line, you can evaluate and extrapolate what the internal portions look like. This will allow you to observe and assess the condition of the line with just a glance. From your assessment, you can then decide when it will become time to replace the stay well before any form of failure occurs.

Dyneema is UV stable, and can live out its entire service life without any protection from the sun. The line will break down a little and will lose some of its original strength, but when sized appropriately, it will not suffer enough loss of strength to merit concern. The advantage of an uncovered Dyneema stay is limited to the ease of inspection; if there is any issue presenting itself, it will be visible just by giving the stay a look. Chafe will appear as a fuzziness on the surface, and from the surface appearance you can estimate the internal condition of the stay. All of this lets you evaluate the condition of the line and estimate when it will need replacement.

Having an uncovered Dyneema stay might sound nice, but covering it has merits as well. Areas where the stay will receive a lot of chafe, be it on the lower portion of the shrouds where the sheets rub or on the headstay where the hanks are present, will benefit greatly by having a chafe sleeve installed over the Dyneema. The chafe sleeve is made of Dyneema but it is a sacrificial piece that can be easily replaced if it becomes too damaged over time without affecting the rest of the stay. They also provide some UV protection to further help the stay have a long life in the sun.

Marine grade lines are much more expensive than the double braid line that you can get at your local hardware store. Some might feel compelled to "save some money" by buying hardware store rope in the correct size and using that to rig their boat. This will work well when it is new, but the inexpensive rope is not able to resist the UV radiation as well as a proper marine grade rope and they will fade and turn to dust in short order. If you feel compelled to save some money, let me inform you that it will not be cheaper in the long run. If you are looking at buying a boat and see what looks like hardware store rope, or the pattern on the cover of the rope doesn't match those of a major producer of marine ropes, beware as you will need to replace all of these deficient ropes promptly! As a side note, if the previous owner used non-marine grade ropes, they probably also used non-marine grade materials on the rest of the boat as well, so you will be looking at replacing everything the previous owner had done to get the boat back to being ship shape. Is it really worth the headache and hassle? It might be if you can identify these deficiencies and bring the price down appropriately!

TEMPERATURE

Air temperature is not something that most sailors would think of as having an effect on their rigging. While for the grand majority of sailors out there, this is the case, as their steel wire rigging and aluminum spar expand and contract at roughly the same rate as the temperature fluctuates, and therefore never need to worry about it; however, those with synthetic standing rigging need to keep a close eye on the temperature, as Dyneema expands as it cools while the aluminum spar will contract as it cools. This compounds on itself as temperatures fall to generate very loose rigging and as temperatures increase, very tight rigging.

If all of your sailing is performed around the same temperature, or if you live in an area where the temperature doesn't vary much throughout the year, then you will probably never need to worry yourself with the temperature and the effects it has on your synthetic standing rigging. If you sail at higher latitudes and in all seasons, then suddenly this will become a very important point for you to pay attention to.

In chapter 10, we briefly looked at which rig designs work well with synthetic standing rigging and which designs can work but will likely be more trouble than it's worth and would benefit from staying with steel wire rigging; now we will discuss why!

Rigs with aft swept spreaders require that the cap shrouds and intermediate stays be tightened to 25% of the breaking strength of a specified steel wire size. The aft sweep of the spreaders takes the compressive force from the shroud and uses it to push the middle section of the mast forward. This force causes the mast to bend, and the greater the force on the sails, and therefore on the windward rigging, the more the mast will bend forward, which in turn will cause the mainsail to flatten, automatically depowering the sail and allowing you to safely carry more sail in higher wind conditions. These rigs are incredibly tension dependent and if the tension on the stay isn't appropriate, the mast will not be pushed forward and the right amount of bend will not be generated. If one of these boats had synthetic standing rigging, every time the air temperature changed, the rigging would need to be adjusted as well! If you have ever sailed through a severe cold front, then you know how quickly the air temperature can drop and how strong those winds can be. The thought of having to worry about how the rig tension has been lost during that situation is enough for me to recommend against using synthetic rigging on any boat with aft swept spreaders.

Rigs with in-line spreaders do not rely as heavily on rig tension for their support. This does not mean that rig tension is irrelevant to them as they still need to have tension on their stays, but as the temperature varies, so will the length proportionately on all the stays. They will all expand and contract at the same rate allowing you a much wider window of appropriate rig tension as temperatures change.

For example, if you like to sail in pleasant weather that ranges from spring to fall, simply set your rigging at 80°F (26.7°C), then you will still be fine when your rigging tightens up as you sail through the summer and temperatures reach up to 100°F (37.8°C), and as you sail into the fall, you will still be fine as temperatures drop down to 60°F (15.5°C). When you set your rig tension with in-line spreaders, you are pretty much good for a ±20°F (±22°C). If you sail in a colder climate, simply tuning your rigging roughly in the middle of your desired temperature range will suffice to keep your rigging in good standing as the seasons change (so long as your desired temperature range is roughly 20°F or °C).

As the temperature drops, all your stays will expand at roughly the same rate and the mast at rest will not have the rig tensions that are prescribed in your manual, but they will take on the appropriate tensions when they are loaded by the sails on the windward side. The mast will lean a little more to leeward until the stays go tight, and everything will proceed as normal from there. If you are sailing on the lower end of your setup temperature range, just keep that in mind and take extra steps to prevent any accidental snaps in the sails. Jibing needs to be executed with caution as the added length to the stays will allow the rig to whip even farther and gain even more speed before crashing to a halt! The shock loads could become too great and in the best-case scenario generate horizontal cracks on your metal fittings, or in a worst-case scenario cause the mast to come crashing down. If sailing below the desired temperature range, it is always prudent to keep the sails reefed. Rather than putting your sails up all the way so that the forces fall on the cap shrouds, which are the longest and therefore suffer the greatest expansion, keeping the head of the sails no higher than your lowers will keep the loads both lower on the mast and lower in force. These reduced loads will fall on the lower shrouds, which are shorter and therefore would not have expanded as much as the taller and longer shrouds on the mast.

As you can see, having a laidback rig that doesn't care about rig tensions as much is much easier to manage with synthetic standing rigging than a high-strung setup with aft swept spreaders that relies on the action of one component to generate a counter action on a different component.

CHAPTER TWELVE
How to Carry Out a Rig Inspection

An ounce of prevention is better than a pound of cure; the old adage holds true with rigging as well. Imagine replacing a part of your rigging while the boat is moored, where you simply remove the old part and install a new replacement part. Compare that to having the mast fall over because that part needed to be replaced and wasn't! You will certainly be replacing the broken part when you replace the mast and all of the associated rigging, but the fix will be a lot more involved.

The trick to preventing failure is to find the faults before they become obvious, and by obvious I mean your mast falls down. Identifying problems and replacing the failed components is the first line of action, but it is also a good idea to take note of "frequent repeat problems" to evaluate if any of these components are being improperly used or if a better component could be used in their place. Imagine that the padeye to a turning block keeps bending, or the padeye keeps leaking no matter how many times you seal it. It might be that the padeye is too small for the job and therefore bends, or, worse, the deck is not strong enough to support the force causing it to flex and break the seal, allowing water to intrude around the fastener holes. It might be a situation where you need to change the padeye for a different size that can take the load, or the deck might need some structural changes to allow it to resist the forces that are being demanded of it. This is all well and good, but you are still only looking at the failing padeye; what you really need to do is look at the entire rig as a whole. What forces are being put on the padeye that holds the turning block? Is the sail whose sheet passes through the turning block too big or is it being flown in strong wind conditions? You need to open your mind to read beyond the problem that you see before your eyes and consider, on a global level, what is the real problem here and how do you fix it for good!

The first step in the process is to collect data on everything. You want to inspect everything at the micro scale, and then evaluate and make decisions at the macro scale. The first part is what people consider to be the rig inspection because that is the part that they observe a rigger performing on their boat. The real part comes later when processing all of this data to form a rig inspection report.

In general, a rig inspection is nothing more than a close analysis of every part of the rigging, beginning with the chainplates at the bottom and working your way up to the top of the mast. I find it best to start with the standing rigging and finish up with the running rigging. I also work from bottom to top because most problems will develop at the bottom and if you find something that indicates that the rigging needs to be replaced at the bottom of the mast, the condition of the rigging aloft is suddenly insignificant because it will get replaced anyways. Once it's broken, you don't need to say it's "broken everywhere," it is just "broken and needs to be replaced." Another reason for starting at the bottom is if the mast is on the brink of falling over, there will likely be giant signs at the bottom, which would prevent you from risking your life while climbing a mast that could fall down with you on it. It is important to note that falling from a mast is incredibly deadly, as a fall from twice your height onto a hard surface earns you an ambulance ride to the local hospital. If you are 6 feet tall, that means that from 12 feet up the mast, you are in big trouble and masts on larger cruising boats are well over 60 feet in height! Finding that the mast is unsafe at the deck level is much safer than being on the mast as it falls while finding problems aloft.

The first steps of a rig inspection are to give every single component a very close inspection, looking for signs of corrosion, discoloration, flaking paint (if any is present), deformation, crack formation, or any sign of damage to the components. The second step is to look at the way and position of the installed components. Are the pins oriented with the head up and legs down? Are the legs spread appropriately? Are the clevis pins the appropriate length for the toggles they secure? Moving up a little bit, you will come to the turnbuckles and once again look for any signs of deterioration as well as the way they are set. Is the turnbuckle tightened down to the point that it is two blocked or is it opened all the way up? This will tell you if the wire is too long or too short. The risk here is that the screw doesn't have enough threads engaged in the turnbuckle and is at risk of pulling free. The other issue is that there might not be enough space for a pin to hold everything in place and prevent the turnbuckle body from rotating.

The last thing to look at with turnbuckles is the situation of the cotter pins. The purpose of the cotter pin is to stop the body from rotating around the screws and thus release the tension on the stay. To accomplish this, the cotter pin needs to stick out to the side, perfectly poised to tear sails, sheets, and shins as they rub against it! To prevent bodily harm for all components on the boat, most people bend the legs over far enough that they are no longer a threat. This blunted approach to cotter pin legs is nice for the health and well-being of everyone on board, but it will no longer prevent the turnbuckle from spinning and undoing the rig tune that you once had. The cotter pin still serves a purpose of stopping the turnbuckle from unscrewing completely and letting the end of the stay go free into the wind!

HOW TO CARRY OUT A RIG INSPECTION

The cotter pin dilemma, to bend or to leave straight, is not a very difficult problem to solve. The issue is that most people are going about it the wrong way. The solution is simply to not use a cotter pin, use seizing wire!

Seizing wire can be wrapped through the hole and around the side of the turnbuckle body. This will prevent rotation while also allowing easy inspection and it has no sharp exposed edges to tear flesh or textile. Seizing wire is also easily attained on a sailboat, as every boat should carry a small roll.

An alternative that I have seen but never tested myself is to install nuts on the screws above and below the turnbuckle to lock it up, similar to when you tighten two nuts onto each other.

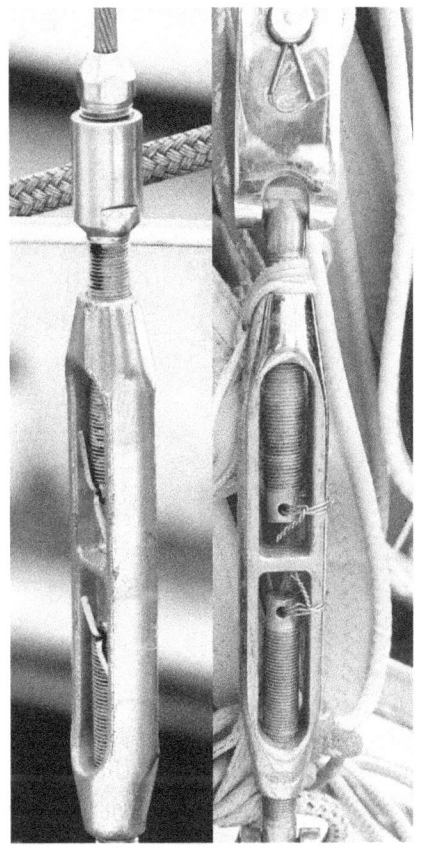

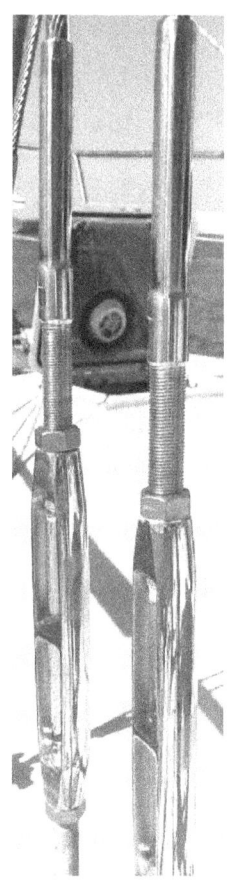

Before we get started, it is important to note the correct way to bend a cotter pin for sailboats, and it is not how they do it on airplanes, because sailboats do not take to the sky with a powerful engine pulling them along. The correct way is to bend the legs 10° each, so that the V that forms between them is 20°. Bending them more than this makes them difficult to safely remove when needed and bending them less than this can let them fall out.

In the aviation industry, cotter pins are bent all the way around so that the legs are up by the head. This is the correct way to bend cotter pins on airplanes, but not on sailboats. Gravity needs to also be taken into consideration when looking at rigging components. "Up" on a sailboat is a relative term, as the boat heels over, up can become sideways, but for all intents and purposes, up will be thought of while the boat is at rest and also the windward side when heeled over. Components should be assembled in a way that gravity holds things in place, and cotter pins simply keep them in place when up is not up. The head of a clevis pin should be higher than the side that has the cotter pin. The head of the cotter pin should be above the legs of the cotter pin. If it's not, it would behoove you to rotate the clevis pin so that the cotter pin is in the correct orientation.

HOW TO CARRY OUT A RIG INSPECTION

The only time that there is any discretion involved is when the clevis pin mounts perfectly horizontal and there is no "up" side at rest. In these cases, the head should be on the outboard side so that when the boat is heeled over and the clevis pin is in use, the head is up and if the small cotter pin falls out, hopefully it will end up on the deck, where it can be found and reinstalled, but the clevis pin can't fall out while it is supporting the mast on that tack.

Below is a worksheet that I use to conduct rigging inspections. It starts at the bottom and works its way to the top of the mast. By filling out the worksheet, your attention will be drawn to each component and that will make you give it a proper look, which is all you need to do to find most problems in your rigging.

RIGGING INSPECTION CHECKLIST

MAST AND BOOM

Mainmast
  Material
    Aluminum
    Wood
    Steel
    Carbon fiber
  Brand: _____
  Mast Height: _____
  Number of Spreaders: _____
  Spreader Orientation
    Aft swept
    In-line
    Forward jumper strut
  Boom Length: _____
  Reefing System
    In-mast furling
    In-boom furling
    Slab reefing
      Tack horn and clew lines
      Single line reefing
      Double line reefing
      Number of reef points
        1
        2
        3

Condition
  Poor condition
  Fair condition
  Good condition
  Excellent condition
Corrosion
  None
  Mild
  Moderate
  Severe
Cracks
  None
  Horizontal
    Location: _____
  Vertical
    Location: _____
Deformation
  Location: _____
  Severity
    Mild
    Moderate
    Severe

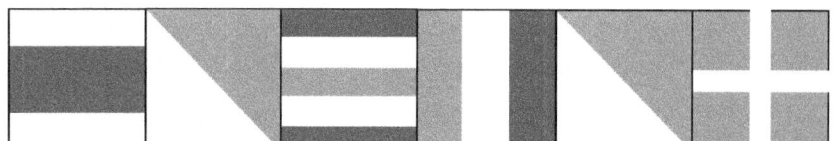

STANDING RIGGING

For each individual component of the stay and on each stay, you want to check the following criteria:

Condition
    Poor condition
    Fair condition
    Good condition
    Excellent condition
Corrosion
    None
    Mild
    Moderate
    Severe
Cracks
    None
    Horizontal
      Location: _____
    Vertical
      Location: _____
Deformation:
    Location: _____
    Severity
      Mild
      Moderate
      Severe

On these parts, you want to also check these additional points:

For clevis and cotter pins
Orientation
    Gravity will hold in
    Gravity will cause it to fall out

For steel stays
Corrosion
    None
    Mild
    Moderate
    Severe
    Candy Cane
Fracture
    Cracked wire
    Raised wire
Swage terminators
    Corrosion coming out of swage
    Corrosion on the swage
    Sign that wire has slipped out of the swage
    Bend in the swage
    Vertical fracture
    Horizontal fracture
    Condition of other end of the terminator
Compression terminators
    Corrosion coming out of the terminator
    Corrosion on the terminator

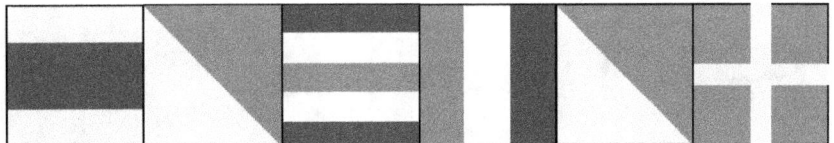

STANDING RIGGING (continued)

　　Sign that the wire has slipped out of the fitting
　　Cracks in the other end of the terminator
Turnbuckles
　　Presence of retaining system in the bottle screws
　　Functionality of the retaining system in the bottle screws
　　Corrosion
　　Cracks
　　Deformation

For synthetic stays
Chafe
Cuts
Condition and length of the taper at each splice
Signs that the splice is pulling apart

For furlers:
Condition of the furling line
Fairness of the furling line
Ease of rotation
　　Spins freely
　　Spins with force applied
　　Resists spinning
　　Seized up
Contaminants that fall out of the furler when you spin it
　　None
　　Salt
　　White powder
　　Dirt
　　Sand
Sound when spinning
　　Free spinning bearing sound
　　Grinding
　　Screeching
　　Grating

For each stay
Condition of:
　　Chainplate
　　Clevis pin
　　Toggle
　　Turnbuckle or deadeye system
　　Toggle
　　Lower terminator
　　Stay
　　Upper terminator
　　Toggle
　　Tang and associated clevis pins

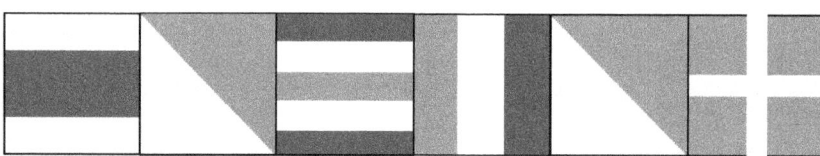

RUNNING RIGGING

For each individual line of the running rigging, you want to check the following criteria:

Condition
  Poor condition
  Fair condition
  Good condition
  Excellent condition
Corrosion of shackles
  None
  Mild
  Moderate
  Severe
Cracks in shackles
  None

Horizontal
  Location: _____
Vertical
  Location: _____
Deformation of shackles:
  Location: _____
  Severity
    Mild
    Moderate
    Severe
Signs of deterioration of rope
  UV degradation
  Chafe
  Overload
  Splice pulling apart

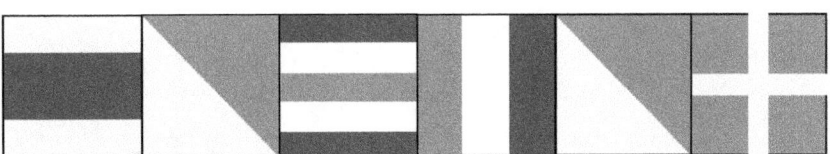

By keeping a close eye on all the components of your rig individually, you can ensure that the rig as a whole will be protected because you will spot the problem before it becomes a catastrophe. As the old adage says, a chain is only as strong as its weakest link. Each stay is the equivalent of a chain, and each component is the individual link. Some are small, like a clevis pin, and some are much longer, like the wire or fiber of the stay, but they all comprise each step of the way linking the boat to the mast. If you maintain each one in optimal condition, then your rigging will keep going strong.

As the rigging ages, it will no longer all look like it was manufactured and installed yesterday. For this reason it is important to be able to judge how much life is left in a rig to predict when the rigging will need to be replaced. As a general rule of thumb for steel rigging, you should replace the rigging every ten years. This means that at a minimum, every year of the rigging's life is 10% of the service life as well. This means that if a rig is two years old and looks like it's still brand new, then it will have 80% of its life left. If a rig is starting to show signs of corrosion and is seven years old, then it has about 30% of its life left. If the corrosion seems to be more advanced, it will further reduce the amount of service life left in the rigging.

The ten-year benchmark is a rule of thumb, and as with every rule, there are always exceptions. If a boat lives closer to the equator in the tropics, where temperatures and winds are higher, the rigging will appear to age prematurely. The Caribbean is a wonderful example to see this advanced aging occur. Steel rigging will begin to rust in the first few years of its life and these boats rarely make it all the way to ten years with the same rigging. Along the way, some stays may experience severe corrosion and necessitate an early replacement. These early replaced stays will be easy to identify because they will be located in high splash zones and yet look newer than any other stay on the boat.

One of my favorite examples was a catamaran that had four-year-old rigging that was badly corroded. The swages had moderate rust pouring out of them and staining everything below them. While the rigging should only be 40% of the way through its life, it looked more like it was 60–70% of the way through. All the rigging had a general dull appearance until you got to the martingale stay at the bow. This stay was so lustrous that it was luminous in the noon sun! The boat was located in the mid-Atlantic at the time of the inspection but the profound state of corrosion prompted me to ask where the boat had been. The owner told me the boat had lived its whole life in Puerto Rico, where the spray from the salty waves and heat from the tropical sun must have wreaked havoc on the martingale stay to cause them to need to replace it so early. Just because a boat is currently located somewhere doesn't mean that it has only cruised the local waters. Seeing this level of degradation was a stark reminder that sailing in Paradise for you might be sailing in Hell for your rigging.

The opposite effect is also noted on boats sailing exclusively in the Great Lakes of the United States. These boats seem to be frozen in time, where the rigging never ages and you will frequently encounter boats with rigging that is pushing 30 or 40 years old, but still looks like it was installed yesterday! These boats only sail for a few months out of the year in a freshwater pool. During the off season, the masts are unstepped for the long winter storage where they relax and await the following short sailing season. This constant removal and installation also works all of the parts, which keeps them fresh and in good working order. Should any part become distorted or hard to work, it will get replaced, which keeps everything else working well in the system. While these boats live in a happy wonderland, if they ever leave these magical waters all of their age will catch up to them in an instant!

I had the opportunity to experience one such rig firsthand. This boat had sailed Lake Michigan for the past years with the same rigging. The owner finally retired and set sail towards the Bahamas. He sailed down the Hudson and exited through New York City, where he then re-entered the Delaware Bay and into the Chesapeake Bay, where I encountered his boat. His rigging had only been exposed to the outside world for about two weeks at this point but his rigging was pouring rust out of every place imaginable. The wires were beginning to flake rust off of them in a way I have never seen before! When the owner told me that the rigging was 40 years old, I believed him but then questioned his judgment to embark on a long cruising voyage with his rigging in such a state of decay. He showed me a picture of his boat from three weeks prior and it looked like it was brand new, yet the rigging I stood before looked rough and rusty. It was an incredible transformation that required a full rerig to bring the boat up to snuff so that he could continue his cruising dream.

To judge the overall condition of the rigging, I recommend considering a few factors: age, corrosion, slipping, cracks, and deformation. Age is linear for steel rigging, where every year is 10% of its service life.

Year	1	2	3	4	5	6	7	8	9	10
Percent Left	90%	80%	70%	60%	50%	40%	30%	20%	10%	0%

Corrosion falls into five categories:

Corrosion	Percent Left
High Luster	100%
Dull Luster	80%
Light Corrosion	50%
Moderate Corrosion	20%
Heavy Corrosion	0%

Slipping and deformation are signs that it is time to replace that individual component immediately, but that doesn't condemn the rest of the rig yet, just that individual stay. If there is a crack that develops on any stay, though, and all the other stays look to be in the same state based on corrosion, then the entire rig should be replaced because it is safe to assume that the other stays have cracks that just haven't been found yet.

Using these tables, you can effectively judge the remaining life of the rigging based on time and corrosion, whichever ages the rig further to make your assessment on how soon the standing rigging needs to be replaced.

While the standing rigging is usually what first springs to mind when people talk about the rigging on a boat, the running rigging is just as important a player in the grand scheme of things. Without the running rigging, you would just have a mast in the air with sails on the deck! Running rigging is the vital piece that makes a sailboat sail and if this were to fail, everything would come to a crashing halt just the same.

Using the checklist above, you can look over every halyard, outhaul, downhaul, sheet, and car adjuster to make sure that the running rigging is up to the task at hand! Running rigging should also be replaced on a ten-year schedule as well, allowing you to reuse the ten-year table to determine the amount of life still left in the line. A supplemental method to evaluate the amount of life left in running rigging is to use a more visual and tactile method of evaluation as outlined in the table below.

Rope Condition	Percent Left
Shiny and clean surface, looks new	100%
Dirty cover but still smooth surface	80%
Surface is starting to fuzz	50%
Chafe spots appearing more frequently	10%
Areas that get thinner rapidly	0%
Any herniation of the core through the cover	0%

A rope is much like a fish: If it looks fresh and smells fresh, it is fresh; if it's starting to look a little worn, it is. New ropes have a wonderful look to them that is easily distinguishable from a "cleaned rope" as the yarns of the cover appear absolutely perfect.

As the rope starts to age, dirt will collect in the cover and go making its way into the core where the little grains will chafe the rope internally. These broken strands will then fray out and appear as a slight fuzziness on the surface of the rope. Cleaning a dirty rope will only aggravate the yarns of the cover and make the rope look cleaner but still fuzzy. A rope that looks like this is starting to age but still has a long life ahead of it. Depending on the level of fuzz, these ropes have anywhere between 50–80% of their service life left in them.

When the rope becomes more worn, its resistance to chafe begins to fade away and simply rubbing against a smooth surface can destroy the cover and cause an area of severe chafe rather quickly. To extend the service life of these ropes, some sailors will patch these parts, deferring the cost of replacement a little longer. These ropes have about 20% of their life left in them and will be in need of replacement soon. In high strain situations, such as sheets, halyards, and dock lines, these lines need to be replaced if they are going to be used in any rigorous situation as they will likely fail during their next serious storm. Lesser stressed lines, such as downhauls, can continue their service but know that their end is very near.

The end has arrived for any rope that has a herniated core, as in the core is bulging out the side of the cover, or a rope that gets very thin in a section. Both of these situations are caused by a failure of the fibers in the core and the cover is now taking the load. When the rope gets very thin in a particular region, that means that the core snapped inside the cover and the cover is actually doing all of the work. Hernias appear for a variety of reasons, but they tend to allude to a partial failure of the core. If some of the strands break, the remaining strands will pull tightly and that will cause the shape of the core to distort, which can then push the bulging parts out through the yarns of the cover. It can also be caused by sun damage on the cover causing it to shrink dramatically and pushing the core to the side as it tries to contract. If the cover is this far damaged, we can assume that it won't do a great job of protecting the core, which is why it should also be discarded and replaced. In either situation, when you see the core blebbing out of the side of your cover, know that the rope is finished and needs immediate replacement.

Using the table above, you can estimate the amount of life left in your running rigging and make determinations accordingly to fit the needs of your boat. If you are a fair-weather sailor who only sails around the harbor, you can let your running rigging get a little closer to the end of its service life before replacement. If you are a bluewater sailor, crossing oceans and battling all the weather conditions that exist on your month-long passage between continents, then it would be prudent to make sure that your running rigging has plenty of life left in it before setting sail as there are no chandleries on the high seas!

CHAPTER THIRTEEN
WHEN TO REPLACE

IDENTIFYING THE CONDITION OF EACH PART OF YOUR RIGGING INDIVIDUALLY IS A very important step, but you shouldn't fall into the trap of treating your rigging as a series of individual components. Imagine that you notice a bit of slippage in a lower compression fitting terminal on your port aft lower, obviously, the correct thing to do would be to open the fitting and replace the failed cone, that way it will hold the wire appropriately. This sounds prudent, but if you simply replace each individual component as you go, you might not notice that everything around it is also starting to show its age.

Steel rigging will crack, corrode, slip, and distort. Synthetic rigging will creep, and chafe. Running rigging will chafe, change diameter, and eventually snap in two. All of these issues are signs that the parts are aging through their service life.

There are two steps to looking at your rigging: The first is to collect data on every individual component; the second is to think about what you have found and see how it relates to the system as a whole. If one stay or halyard is in really bad condition and the rest of the rigging looks like new, the answer is easy to see, that one part of the rig needs to be replaced. If that same stay or halyard is looking really worn out, and the rest of the rigging is also approaching the 20–30% remaining life mark, then it might be better to simply replace all the rigging.

Full rerigs are obviously more involved, as everything is removed and replaced with new. This is different from a partial rerig where only select parts are replaced. Picture a monohull with five-year-old standing rigging that looks rather aged but has a brand-new furler and headstay. This couple just bought the boat earlier in the season and they are living aboard while they finish out their careers for the next three years. Once they retire, they are going to sail to the Bahamas via the ICW on the East Coast of the United States. For the next several years, they will only be doing weekend sails around the local inland waterway, and only during fair weather. Over the next few years they will gradually outfit the boat for cruising to avoid breaking the bank right as they stop working. Looking at the running rigging, you notice that the lazy jacks are really tight and the remnant of the topping lift is tangled around the end of the boom; the main halyard is also really stiff to raise and lower the sail. What would you do in this situation?

The headstay and furler were replaced by the previous owner right before they bought the boat but they are wondering if the entire rig needs to be replaced. The topping lift obviously broke and siphoned itself into the mast where it tangled around the halyard. As you can imagine, the rigging on this boat is failing and needs to be replaced but when should it be replaced?

If you rerig the entire boat now, you are setting them up to have older rigging when they finally go and start to use the boat. This means that their next rerig will occur a few years after they finally get in the groove of cruising and have a more limited budget. A discussion with the owner about the global situation will reveal an excellent compromise. The topping lift is broken and obviously needs to be removed. The main halyard is of similar vintage and since they work in the same area and are made of the same material, it would be prudent to replace the main halyard at the same time as the topping lift, but closely monitor all the other running rigging to see if we can't squeak out a few more years on it. Knowing that the rerig is going to come in a few years allows the couple to set aside funds for the project and spreads the hurt out a little, since they did just buy the boat! In a few years, they will be retired and ready to go cruising, so just before retirement, the rigging can be replaced that would give them brand-new rigging with 100% remaining life in it, as they set off and begin their retirement voyage. This would let them enjoy their new rigging while actually sailing instead of going through the hassle of rerigging only to finally start cruising on aged rigging. The way their current rigging aged indicates that the environment where their boat is kept is harsh on the rigging, so there is no point in putting a new shiny piece of hardware on their boat that's just going to sit and stew in that harshness for the next several years. Worst case, they might need to rerig again right before they leave, which would double their cost for absolutely no benefit.

Looking at the global situation you will recognize not only what but also when things should occur for the betterment of the boat. If you simply replace each single part as it fails, you will always have parts failing and you will always be replacing something on your rigging. Having a firm cutoff at ten years ensures that every decade, every part that was questionable will be renewed so that if a problem has gone unnoticed, it will be taken care of at the ten-year mark.

Think of it another way: Picture a hundred parts all wearing down at a similar rate. When one of them starts to go, the other 99 are close behind. If you are replacing parts individually, then you will find that your to-do list is never ending as these parts will all start to fail around the same time. If you keep at it for long enough, then the

WHEN TO REPLACE

variation in lifespan of each part will spread your maintenance out to the point where you will constantly be replacing parts and never get to enjoy your boat as something will always be broken and needs your attention; let's just ignore reality where that already is the case. No need to add to your boat's endless to-do list and take care of the rigging in one fell swoop so that it will be good to go for years to come.

My personal feelings on when to replace one part versus replacing the entire rig is rather simple. If one part fails before the rigging as a whole has reached 50% of its service life, then I replace that one part. When the rest of the rigging reaches its inevitable end, that new part gets replaced as well, that way the entire rig is of the same age, lest we find ourselves in the scenario above where each stay is of a different age and requiring replacement on a different year. That is, first, too much work, and, second, too much to remember! If the rigging has less than 50% of its life left and one of the stays fails, I would then recommend replacing the entire standing rigging.

Using the tables above, you can determine if the rigging has reached 50% of its life based on time and condition. It is important to note that the standing rigging and running rigging run on their own schedules, and while they will probably overlap with each other, they do not need to occur in the same year. When the boat is new, the running rigging and standing rigging will naturally be of the same age. When the boat reaches ten years of age, so will the standing and running rigging, and therefore both will need to be replaced at the same time. The reality is that the owners of the boat will probably change the sheets and halyards due to color preferences or material choices. Some new boats tend to come with generic and nondescript double braid Dacron or polyester lines that "tick the box" for the dealer to state that they supplied the sailboat with running rigging, but that's about all that type of rope is good for. This kind of rigging will die in a few years and need to be replaced well before the ten-year mark, which is why the running and standing rigging get their own independent countdowns.

So far, the entire discussion has revolved around time, as that is the constant that moves all boats through their age. Most boats spend their entire lives sitting at a berth in a marina, waiting for the few weekends a year when they will get to go sailing. As they pass the days baking in the sun and exposed to the elements, they will gradually decay as the only water that passes by their keel is that of the tidal current. For the few boats that actually go the distance, time is ignored in favor of a more tangible unit: nautical miles. For boats that are sailing far, the standard switches from 10 years to 10 years or 10,000 nautical miles. An ocean crossing will usually rack up about 5,000 nautical miles, which is why it is a good idea to embark on your cruising life with new rigging. Imagine that the standing rigging was three years old and had roughly 70% of its service life left, upon arrival at this magical island, your rigging would have roughly 30% of its service life left, which means that you need to find a place and materials to rerig before your next big voyage.

You're probably thinking to yourself that you have heard of lots of people who circumnavigated over ten years or more and never replaced a single part of their rigging. I have met them and I can tell you that their rigging is long overdue for replacement. These sailors are doing the equivalent of driving their car well past the time it was due for its oil change. What happens to your motor if you drive your car for longer than 10,000 miles? Does your motor lock up and explode as the odometer reaches 10,001? Absolutely not—as long as you still have something that resembles oil in there, the motor will keep spinning and you can keep driving. Around 20,000 miles, your motor is still going strong and you might think that the whole oil change business is a scam that's ripping motorists off for generations. As you reach 50,000 miles on the same oil, you might start hearing some strange sounds and think the car just needs a tune-up. The first mechanic who looks inside your motor will look at you in wonder as they cannot fathom the reasons that drove you to drive so far while neglecting your motor. The oil has turned to sludge and the passageways that used to lubricate your motors internals are now clogged up like my uncle's arteries. The engine is destroyed because this simple maintenance was neglected and now the car is looking at a hefty repair bill to get back on the road!

The same holds true with your standing rigging. If you ignore it, your mast probably won't come crashing down. Corrosion will begin to form and terminators will begin to slip. Cracks will form hidden from sight, all waiting for the moment when the proverbial piece of hay comes crashing down to break the camel's back. While people neglect doing simple maintenance on their car, other people neglect doing maintenance on their boat and most of these people get away with it!

I have met multiple elderly sailors who are proud to tell me that their boat that they bought in their thirties has original everything! Original standing rigging, original sails, original motor, and original running rigging. They seem to view it as a thing of pride, proof that they have maintained their boat at such a high level that nothing has deteriorated on it over the decades. In their defiance to the precautions of known service lifespans, these geezers happily sail away in complete ignorance to their precarious situation. Does this make it something you should imitate? Absolutely not!

The average world cruiser takes about ten years to comfortably circumnavigate the tradewind routes while also enjoying their time in the various warm destinations that their boat has delivered them to. This route is also around 26,000 nautical miles at a minimum, which means that these sailors should theoretically need to rerig at least twice on their voyage around the world, yet almost all of them make it back to their home port with the same rigging they left with. They might have never thought twice about their rigging, as it was good yesterday so it will be good today! I think that these people are just plain lucky. Their rigging, if inspected, is sure to turn up a smorgasbord of problems and the fact that the mast remains standing is in direct defiance to the reality of the world we live in; in other words: luck.

This whole 10-year or 10,000-nautical-mile rule only pertains to steel standing rigging. Good quality running rigging that is sized appropriately to your boat will die because of time and not because of use. The sun with its punishing UV radiation will bombard the fibers of the rope and cause them to break down to the point they will actually break. The harsher the sun that you sail in, the sooner the end will come. With running rigging, an attentive eye and prudent protection from all chafe will allow you to exceed the ten-year mark but this will be borrowed time, and when the debt is called, you will have to pay up!

The first sign of significant chafe caused by mild abrasion is your cue to the fact that the bell has been rung and your rope is done. There is no more repairing at this point; the running rigging now needs replacement. Obvious ways to extend the service life of your line are to protect it from the sun and keep it in a place with good airflow. If the sun doesn't kill it, mildew growing inside of it will. Sheet bags with more ventilation are ideal for this as they will keep your sheet protected from the sun and dry. Stowing your coiled halyard inside the mainsail cover will also protect that line from the sun's UV radiation. These tricks will help you prolong the inevitable, but it won't actually stop it from coming. Keeping an open mind to the fact that the running rigging has reached its age and has finally reached the bitter end of its life will let you save your money by deferring maintenance on your running rigging until just before it is actually needed. As always, you don't want to wait until it breaks because that negates the entire point of inspecting your rigging to catch the problems before they develop into big problems.

Earlier, you read that steel standing rigging needs to be replaced every 10 years or 10,000 nautical miles, and steel standing rigging is what most cruisers are sporting as they navigate the high seas to lands they have yet to step foot on. The secret to going longer distances safely is to not use steel standing rigging, and instead use synthetic standing rigging. If you noticed on the inspection checklist, steel standing rigging had a lot more parts and a lot more problems that needed to be checked for. Synthetic standing rigging is immune to most of these perils. Dyneema is the most commonly used material for synthetic standing rigging on cruising boats.

With synthetic standing rigging, you can sail much farther and longer on the same rigging because the plastic that makes Dyneema is unphased by the marine environment. Plenty of data has been collected by DSM, the manufacturer of Dyneema in the Netherlands, on estimating the remaining life in their ropes. These ropes are used extensively on oil rigs to moor them in place, as well as mooring lines of almost every giant cargo ship that sails around the world. If one of these mooring lines were to break, the cost of lost time and damage to the resources being transported would be great. For this reason, extensive study has been performed by DSM to accurately predict the amount of service life remaining in these ropes. DSM is so kind as to publish these findings and with it you are then able to accurately calculate the service life of the stay. When SK78 Dyneema is loaded to 20% at 16°C of its breaking strength, it will creep approximately 0.5% per year. SK78 ruptures from creep at around 30%, which means that you would have about 15 years of service life from your rigging. As you know from previous chapters, you would never dream of sizing your rigging to be at 20% of its breaking strength, as the creep will be rather dramatic. 0.5% doesn't sound like very much but on a 20-meter-long stay, that would be approximately 0.1m per year, or 10cm (3.9 inches) per year. That is way too much creep for any sailboat, which is why they need to be sized appropriately so that they live in a world where they are loaded at a much lower percentage of their breaking strength. If you undersize your rigging so that it will creep almost 4 inches per year, it would last you about 15 years before it would finally reach its end. If you size your rigging appropriately, it will creep significantly less and therefore last significantly longer!

Dyneema is unaffected by water, salt, chemicals, acids, everything that the marine environment can throw at the rigging, which means that the only real force working against your rigging is the strain it is subjected to. Doing proper calculations before you build your rigging will allow you to select the proper size that will be willfully underworked on your sailboat and will, as a result, last for years beyond the ten-year lifespan of steel rigging.

While your synthetic standing rigging will last a very, very long time, it will eventually reach its end in two ways. The first is the natural progression of its life cycle where it finally creeps to the point that it reaches the third phase of its life cycle and begins to creep at an astonishing rate. This is your sign to immediately procure new Dyneema and get splicing. The second is due to chafe.

In summary, steel standing rigging is replaced every 10 years or 10,000 nautical miles, whichever comes first. Synthetic standing rigging is replaced when it finally reaches the third phase of its life cycle, which if sized appropriately will be well after ten years. Running rigging is replaced every ten years or at the first sign of rope degradation if you are doing bluewater sailing, but you can stretch it out a little if you take really good care of the lines and are staying on inland waterways where chandleries are prevalent to replace the rigging at the first sign of the approaching end, which won't be long past the ten-year mark.

CHAPTER FOURTEEN
Avoid Repeating Your Mistakes

The rigging needs to be replaced—it could be due to age, distance, condition, or specific needs, but the point is: Now is the time to renew the rigging. You could replicate your old rigging but that would replicate any mistakes that could exist in your rigging that you were simply unaware of. What if the rigging that is on there is the wrong size or type of rigging for your sailboat?

Let's start off with choosing the right size of rigging that your boat will need. Many sailors who are going to be sailing offshore will intentionally choose to make their boat less seaworthy because they think it will be better. This mistake is called "oversizing your rigging." The reason they do this is very simple and actually quite logical: thicker rigging is stronger and, therefore, the stays are less likely to break and, therefore, the risk of losing the mast is reduced. Wonderful!

Looking at the table on p. 137, you can see how as the wire size increases, so does the strength. The concept is simple enough, selecting the wire size up from the one that you have might still allow you to connect it to the same hardware.

Wire Size (Inches)	Clevis Pin (Inches)	Wire Size (Metric)
3/32	1/4	3mm
1/8	1/4	4mm
5/32	1/4	
5/32	5/16	4mm
3/16	5/16	5mm
3/16	3/8	5mm
7/32	3/8	6mm
1/4	3/8	6mm

(continued)

THE RIGGING HANDBOOK

Wire Size (Inches)	Clevis Pin (Inches)	Wire Size (Metric)
1/4	1/2	6mm
9/32	1/2	7mm
5/16	1/2	8mm
5/16	5/8	8mm
3/8	5/8	10mm
7/16	3/4	11mm
1/2	3/4	12mm

The table above shows you the sizes available from Hayn for a turnbuckle that has an open jaw with a clevis pin for the chainplate on the bottom and a swage stud on the top to connect to wire. This is the easiest way to visualize how a manufacturer intends you to use their product and the range in wire sizes that correspond to each clevis pin size. If you are uninformed, you might think that this system is set up this way on purpose to make it easier to "improve" your boat. When you replace your rigging, all you have to do is select the next size up in wire and it will probably fit into your current setup. If you have $5/16$-inch wire and $5/8$-inch clevis pins, then it looks like they are practically inviting you to increase your wire size from $5/16$ inch to $3/8$ inch as they all connect within the $5/8$-inch clevis pin system.

Wire Diameter (Inches)	Weight per 1000 feet (in Pounds)	Strength of Wire (304) (in Pounds)	Strength of Wire (316) (in Pounds)
3/64	4.8	299.8	266.8
1/16	8.5	533.5	474
5/64	14	833.3	740.7
3/32	19	1201.5	1067
1/8	35	2134.1	1898.2
9/64	43	2641.1	2347.9
5/32	55	3335.6	2965.2
3/16	77	4801.6	4268.1
7/32	102	6536.6	5811.3
1/4	135	8538.4	7588.2
9/32	170	10804.7	9605.4
5/16	214	13340	11858.5
3/8	314	19210.9	17076.8
7/16	410	26146.6	23240.9
1/2	544	33620.2	29883.4

AVOID REPEATING YOUR MISTAKES

As you can clearly see, increasing the wire diameter just $1/16$ inch, which sounds like nothing, adds almost 100 pounds aloft (per 1,000 feet of wire). While you might be thinking: "It won't take 1,000 feet of wire to rig the boat and at least it will be stronger with just a little weight penalty, it's not like I'm racing." The fault is twofold: First, you are adding weight aloft, which means that more of the ballast is now dedicated to simply keeping that weight up instead of resisting the wind pressure on the sails. This means that you will need to reef earlier because your ballast did not increase proportionally to the increased weight of your standing rigging. Second, the added diameter will cause more windage, which means that when you reef down to your skivvies and are only flying tiny storm sails, the windage from the thicker rigging will still cause you to heel over.

In other words, while attempting to make the boat safer, increasing the rigging wire size one step up actually makes everything worse! It's the same as if you had attached a large weight high up on the mast, it does nothing to improve the sailing and everything to detract from it.

While it might seem like a "no-brainer" to add an extra 6,000 pounds of breaking strength to your rigging by going up a size in our example, you now know why that is a bad idea. Now that we all know why oversizing your steel rigging is wrong, we can focus on how to size your rigging correctly.

If you need to make your rigging as strong as possible and are going to be staying with steel rigging, consider using 304 instead of 316 as 304 has a significantly stronger breaking strength. Using 304 instead of 316 would give you a stronger setup without the weight or windage penalty. 304 is less resistant to corrosion than 316, meaning that your rigging would develop a bit of a rust problem before the ten-year mark rolls around, but this wouldn't be a real concern as you would be replacing your rigging based on miles well before you reached the ten-year mark. This is a simple way to increase the strength of your rigging by trading for corrosion resistance without adding weight aloft and messing everything up.

The correct way to size your rigging is based on what the naval architect who designed your yacht wanted. The naval architect came up with the entire plan for your boat and focused on every aspect of how it would sail and perform. The weight aloft was carefully calculated and that, along with many other factors, such as the designed sail plan, helped the naval architect determine how much ballast the keel needed to have and where the ballast needed to be situated to achieve the desired results. If the original plans are available or if the naval architect is still alive and able to be contacted, it would behoove you to find out what the rigging is supposed to be.

If the original drawings are unable to be attained or the naval architect is no longer with us in this world, then you will need to figure out what size your rigging needs to be and size it accordingly. There are two ways to figure this out, one is a quick and dirty method that puts you in the ballpark, and the other is very precise and will tell you exactly what you need. The easy method is to use the clevis pin hole size to find the range that the wire size is supposed to be. Comparing with a few sister ships or asking an owners association is a good way to narrow down the correct wire size at this point, all without having to do any calculations. If this is not available, then your last option will be to do the calculations yourself as outlined in chapter 10.

By doing the calculations, you can figure out the RM30 of your boat and therefore calculate the required breaking strength of the stay so that you can select the correct wire. Personally, I recommend doing this even if you have all the information presented to you, just to double-check the work of the naval architect. It's not that I am saying you can do their job better, but they design a lot of boats and some get a lot of attention while others look like they are cobbled together out of leftover ideas. Doing the math yourself simply checks their work and also accounts for variations between calculated forces and real-world forces. This is a big issue with older boats that were designed before computers as they didn't always float on their lines or heel over as expected. In the rare event that the rigging of the boat is actually oversized for what it needs it to be, you can benefit from shedding weight aloft by installing standing rigging that is the appropriate size.

When rigging is being replaced because of time or mileage, that is a good indication to you that the rigging setup was appropriate for the boat and the rigging worked well for its entire service life. When the rigging failed early, it is important to identify why it failed and take corrective actions to mitigate the recurrence of this same issue.

Looking beyond the sizing of the components and also at the composition of them is critical as the right sized part might simply fail because it is living in the wrong environment. A freshwater sailboat is going to have less demanding requirements from a corrosion standpoint than an equatorial salty cruiser. Stainless steels, aluminums, titaniums, and bronzes are all available in different alloys and grades to withstand different conditions. Choosing the right material can sometimes be more important than choosing the right part!

The moral of the story is to, first, always look at what was done and figure out why it was done in such a way. Second, identify problem areas and how to fix them. Third, identify how your fix is going to fail and resolve that issue now so that you aren't replacing your own work early! If everything looks like it was fine and simply aged out, then do the math and make sure that it was the right setup for the application, then replace it like for like.

The wire, be it 304 or 316, should be decided on at this point due to the desired properties of the metal. The terminators should be set up using compression fittings to avoid repeating the mistake of using swage fittings, as swage fittings are work hardened and more prone to cracking than a well-made compression fitting. Toggles should be installed at the top and bottom of each stay, and all the clevis pins should be replaced with new ones that are installed in a position where gravity will help hold them in while the boat is sailing.

If you are converting to synthetic standing rigging, then you don't need to worry about repeating old mistakes, as you will be going in a different direction, which is why it is important to make sure you don't create new mistakes in the process. The goal with each rerig is to create a system that will withstand the next service interval. If you are using steel, everything should be built to last at least ten years, but if you are using synthetic, everything needs to last more than ten years.

CHAPTER FIFTEEN
Steel versus Synthetic

In the previous chapter, we focused more on replacing the old rigging with new steel rigging and how you should follow the outlined parameters from the naval architect. When converting to synthetic, you are changing everything for the better, so the original schematics are now only reference points to guide you in the conversion.

Why should you modify the rigging by using synthetics instead of steel, after reading through an entire passage where sailors who "up-sized their rigging" were chastised for deviating from the sacred plan that a naval architect bequeathed unto the seas? Because you are going to be making it better using technology that didn't exist back then.

Material sciences have jumped through time to a future where incredible properties are available without the weight or susceptibility to corrosion. Living in the present day, where these materials are easily attainable, the thought of "should I rerig with synthetic?" should at least cross your mind. The goal of this chapter is to help you identify: When should you use synthetic? Why should you use synthetic? When should you not use synthetic? The worst thing you can do on your boat is install the best tool for the wrong job and the goal here is to help you avoid that time-consuming mistake.

WHEN SHOULD YOU USE SYNTHETIC STANDING RIGGING?

Synthetic standing rigging is an excellent choice for replacing stainless steel because it is unbothered by the marine environment. As we've already discussed, it doesn't absorb water, doesn't corrode, and acids have no effect on it. In other words, where steel is constantly being eaten away by the hot acidic environment of salty water in the tropical sun, Dyneema is unphased. Using synthetic standing rigging seems like a no-brainer as it is definitely up to the task and also won't suffer from the environment that it will be installed in.

The ideal application is a sailboat with a single set of in-line spreaders or multiple spreaders that have continuous stays. You want all your deadeyes or turnbuckles to be at the deck so that adjustments are easy to carry out, especially in the beginning. You want to do this on a boat with in-line spreaders because they need less rig tension at rest, so when the temperature fluctuates, the rig will not suffer from the changes in rig tension.

The ideal sailing area is one where your sailing season has a pretty uniform temperature. This could be in the tropics where it is hot all year long, or in a cooler high latitude location where you will only be sailing at a specific time of year when the temperature is at a consistent level. No more than a 40°F (22°C) temperature range is ideal, as this allows you to set your rigging in the middle of that range and then have +/-20°F (+/-11°C) temperatures from that starting point.

If you are in high latitude areas where the winters get cold, your rigging will go very slack in the coldest of months, which is why this next point is important. You either need to have a keel-stepped mast or plan on unstepping your deck-stepped mast at the end of the sailing season. The rigging will go very slack as temperatures drop, and with a keel-stepped mast the mast will remain supported by the deck hull junction while in the boat's berth. When the sailing season arrives again and the air warms up, the rigging will tighten right back to where it used to be and your mast will be perfectly in tune for next year's sailing. If you have a deck-stepped mast and you won't be unstepping it for the winter, you will then need to go out to your boat and tighten the rigging as the temperatures drop. This will take up the slack and keep the mast supported but you must be close to your boat and willing to go out there on a whim if there is any chance of a warm day during the winter. This warm day will bring a nice shortening of the stays and that would greatly over tension your rigging, possibly breaking something! For this reason, it is best to unstep a deck-stepped mast with synthetic rigging so that there is no stress or worry caused by the temperatures as they fluctuate through the seasons.

If your deck-stepped boat will remain in the tropics where it is hot all year round, then synthetic standing rigging will be a wonderful setup for you to use. It will resist all the other ailments that stainless steel rigging will suffer from and provide you with a sturdy and resilient setup in paradise.

WHY SHOULD YOU USE SYNTHETIC STANDING RIGGING?

The reasons to use synthetic over stainless steel are rather simple: It is a better material for the marine environment. It can't corrode, it weighs less, and it is stronger. Dyneema is made of polyethylene, which is a plastic, and as those old commercials in the nineties would say, "Plastic makes it possible."

STEEL VERSUS SYNTHETIC

When looking at the checklist for a rig inspection, you might have noticed all the aspects that need to be inspected on a steel rig because it can have many types of failures, all of which are miniscule and hard to identify at first glance. The checklist for synthetic is much shorter as you are just checking for chafe. If the line is fuzzy, you know you need to address it. Fuzzy spots on a line are much easier to spot, even at a distance, compared to searching every square millimeter of a metal fitting for a microscopic crack.

By making it easy to identify problems early, it enables you to safely judge the condition of your rig before each sail. All you need to do is look up and see if anything looks fuzzy. If everything looks like it should at first glance, you can then proceed to raise your sails and have a great adventure! These more frequent inspections also will reveal an issue earlier where it is easier to address and repair, rather than hoping everything is fine with your steel rigging as you wait until next year's thorough inspection.

The fact that synthetic standing rigging is unphased by the marine environment should speak volumes for the positive reasons to use it over steel. Steel rigging had to undergo metallurgical alchemy to make it able to survive just ten years on the ocean. Rust has always been the biggest hurdle to overcome and the marine environment is the perfect place for rust to happen. Changing the alloy composition over the decades has enabled the metal to postpone its inevitable death, but it is still fighting an uphill battle. The name says it all: stain-less steel, because it is steel that will stain less than regular steel, but the name isn't stain-proof steel. The reason ten years is the standard service life for rigging is simply because steel rigging is the standard and as a result, the standard service life revolves around the service life for steel rigging. Back in the days of hemp and manila rigging, the service life was much shorter and the replacement interval was set accordingly. The whole reason that steel became the rigging material of choice is because it was better, both in strength and durability, to a natural fiber rope. Hemp and manila do not do well in the marine environment but they were pretty much all that was available to make rope. If they got wet, which you would assume would happen very quickly on a boat, they would rot and need to be repaired or replaced. Tall ships had to carry massive amounts of extra rope on hand to replace the stays as they failed.

The other advantage that steel has over natural rope is the strength. Looking at old tall ships, you will notice that the shrouds are literal shrouds! There can easily be more than five shrouds veiling each side of the mast. This is because rope rigging was not that strong so the shrouds needed to be sistered to make the combination of all of them strong enough to resist the demands that would be placed on them while sailing. It also allowed for the replacement of an individual stay with the mast up and in service as the load could temporarily be supported by the other stays. If a stay failed because its condition was not properly determined at the last inspection, the mast would usually remain standing while a new stay could be fashioned in its place. Imagine the joy of a naval architect when a new material comes to market that can replace all those stays with one single stay.

If naval architects were so quick to jump ship from natural fiber rope to steel, don't you think they would have been just as quick to jump at Dyneema? Looking at old tall ships, you can easily see that the rigging setup was drastically different from the way steel rigging is set up, so you wouldn't simply switch the material and maintain the same setup, expecting it to work well.

The same holds true for synthetic rigging. Modern vessels that are designed to have synthetic rigging do not have the same setup as those with steel rigging. Boats that do best when converting are those that are originally designed for steel rigging but the setup is the same as it would be for synthetic rigging. Linked rigs with aft swept spreaders, for example, work great with steel rigging but would be a nightmare with synthetic rigging.

Boats designed with Dyneema rigging in mind tend to have either one or no spreaders on the mast, and the stays are adjusted at the deck. This is because there will be some adjustment that is necessary while the rigging settles in. This is not to say that it can't be done on a linked rig setup, but it would just be very tedious to have to climb the mast every time you need to tighten the rigging as it creeps.

STEEL VERSUS SYNTHETIC

If you have a keel-stepped mast with in-line spreaders and a continuous rig, converting to synthetic standing rigging would be very straightforward. The benefits would be easier inspectability, longer service life, easier fabrication and replacement in remote areas, and less weight aloft. These benefits are compounded upon if you plan on cruising in tropical climates where steel rigging will surely suffer an early death.

WHEN SHOULD YOU NOT USE SYNTHETIC STANDING RIGGING?

The worst thing you can do on your boat is install the best tool for the wrong job. Synthetic standing rigging might be the best tool, but if your boat is the wrong job, then it will be horrible. When you should *not* use synthetic standing rigging is a very simple question to answer: when you would be better off with steel rigging.

There are a few key points that you need to ask yourself about first. Do I enjoy sailing in a blizzard? Are my spreaders swept aft? Is my mast deck stepped? Do I ignore small details on my boat? If you answered yes to any of these questions, then the rest of this chapter is going to be very important for you to help you decide if synthetic rigging is right for you.

Do I enjoy sailing in a blizzard? Weather is a very important deciding factor and, honestly, the first factor that you should look at when determining if you should use steel or synthetic. As Dyneema cools it expands and that makes the entire stay longer. Aluminum, which is the most common mast material, shrinks when it cools. This means that your stay gets longer as your mast shrinks and the result is very loose rigging. If you sail with slack rigging, the mast needs to lean over farther before the rigging will support it and if the mast leans too far it can buckle and collapse, not to mention the fact that your sail shape is dependent on the position of your mast and tension of your rigging; in other words, your sail shape would be garbage until your mast comes crashing down. This is why synthetic rigging performs best within a 40°F (22°C) window, being +/-20°F (+/-11°C) from the temperature that you set it. If you tuned your rigging at 80°F, then you are good for anything between 60°F and 100°F. If you are sailing in cooler weather, then you simply need to tune your rigging in the cooler weather. Say you tuned your rigging at 50°F, then you would be fine anywhere from 30°F to 70°F, but if you then sail into warmer weather, your rigging will continue to tighten to the point that something will break.

On a side note, this is why it is imperative that you keep your clevis pins when converting to synthetic standing rigging because they will be the weakest point in your rigging and will act as a fuse to protect the rest of your boat. If you set your rigging in the winter and then forget to adjust it as spring arrives, by summer your rigging will have contracted so much that it will cause something to break! If you manufactured custom chainplates with nifty holes in them for the lashings to pass through and fashioned a cool eye splice that wraps around your mast, completely eliminating any weak points caused by the clevis pins, then something is going to break and it will be costly when it does! Keeping the clevis pins at the chainplate and mast provides you with two "fuses" that will break when the load in the system exceeds the breaking strength of the pin and that will prevent any further damage to your boat and spar.

If you like sailing in blizzards or cold weather in general, then you must be willing to adjust your rigging in that frigid climate so that your rigging will be properly tensioned for the temperature range that you will find yourself in. This also relates to the fourth question originally posed.

Are my spreaders swept aft? Aft swept spreaders, as you know, require higher rig pretensions to induce the forward bend in the mast that the mainsail depends on to maintain its proper shape. Aft swept spreaders make sailboats perform wonderfully, but in the same way that a high-performance engine requires a specific oil weight, temperature, and pressure to lubricate everything effectively, aft swept spreaders demand a precise amount of tension be applied to them so that everything else follows like clockwork. If the tension on the shrouds is not exactly 25% of the steel stays, breaking strength, then the mast will be out of tune and will not bend as it is designed to. The problem is compounded even further with fractional rig setups, as the leverage between the headstay and backstay can twist the mast into an inversion and cause the instant destruction of your spar.

Is my mast deck stepped? This question really matters if you are sailing in climates where the temperature will fluctuate more than 40°F (22°C) as your rigging will become very slack and your mast will not be properly supported. Deck-stepped masts derive all of their support from the standing rigging, which means that if the standing rigging goes slack, the mast will lean and could damage the heel of the mast where it stands on the mast step. If you plan on cruising the Northwest Passage with your deck-stepped boat, then Dyneema will be a massive hindrance in your setup. If you are cruising on your deck-stepped sailboat in the tropics where the temperatures range from the mid-70s°F during the dead of winter and up to the high 90s°F during the peak of summer, suddenly synthetic rigging will become an asset on your boat as it will do its thing and you won't have to think about it between your rig inspections. In short, deck-stepped masts rely on the rigging to hold the mast up. If the weather gets really cold, the rigging won't be there to do its job.

Do I ignore small details on my boat? This is the most important question of all and each previous question is dependent on this answer. If you enjoy sailing in really cold weather, then you will simply need to adjust the rigging prior to going sailing, and then adjust it again prior to the eventual warm weather to come with the following summer. If you ignore small details, then you might not care to tighten the rigging just a little bit because "what difference does a few millimeters make for the rig tension?" Sailing with the rigging too loose can cause a host of problems. First would be the improper sail shape, which would lead to abysmal sail performance. Second would be the undue stress on the spar as it would be under load and out of column. The mast is only strong when the supporting part is under it. When the mast bends over too far, it will then become unsupported, which can lead to buckling and collapse of the mast! If you set your rigging in the summer and decide to go for a New Year's sail, and you live up north where it gets really cold in January, your rigging will be slack enough to jump rope with. Going full sail with rigging that slack is the same as going full sail without any rigging attached. The mast is going to get hurt!

Paying attention to small details doesn't mean working harder in cold weather, as it simply means being aware of the situation and working within the confines of what is allowed. On my own sailboat, I have a single set of in-line spreaders. Being a cutter, I also have check stays, or diagonal stays, depending on how you want to view them, that oppose the inner forestay three-fourths of the way up the mast. I set our rigging at 80°F (26°C), which means that our safe temperature range is from 60°F (15°C) to 100°F (38°C). A little above or below isn't a huge issue, but when it leaves that range, I pay even closer attention to the forces I am putting on the rig. If it is 50°F (10°C), I simply won't put up the headsail, as this would load the top of the mast, which is the part that is going to have the most flex as the cap shrouds are the longest and will expand the most. I keep the loads at or below the inner forestay and check stays by raising only the staysail and the main with two reefs in it. This brings the load down from the longest stays to the portion that is supported just below. If it is 40°F (4°C), I will only fly the staysail and trysail, and only sail on a broad reach as we are now 40°F (23°C) below the original temperature where the rigging was set, and only in calm winds. When I have found myself needing to use the sails in weather down to 30°F (-1°C), where the rigging is incredibly slack, I only set the trysail in light winds with the head of the trysail positioned at the spreaders so that any load is being well supported by the mast/deck junction while making sure that the lowers don't get tight, as this would indicate that the mast is moving around enough to make the rigging come into function. I know that the rigging is ineffective at this temperature and that means that I was supporting the entire sail plan on the equivalent of someone holding a boat hook with a bed sheet tied to it up into the air. It is very flimsy, but if the conditions are calm enough, it will make you move slowly through the water.

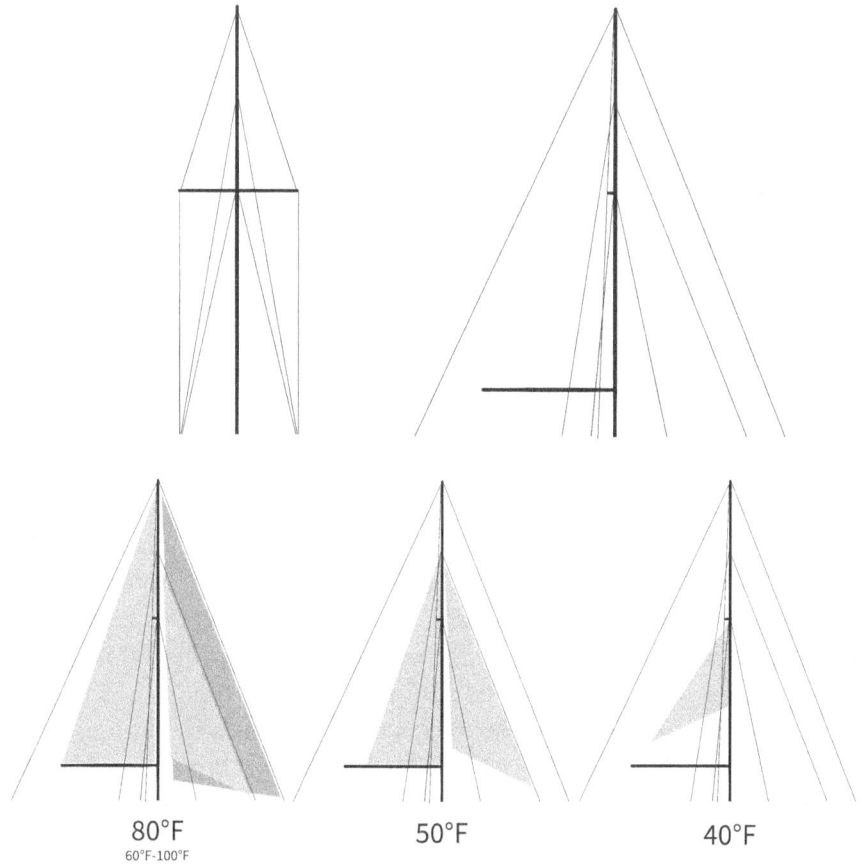

80°F	50°F	40°F
60°F-100°F		

Paying attention to small details like the temperature does not mean you have to freeze your tail off to keep the rig tension at its ideal setting; it simply means you are aware of the situation and act accordingly. If it were really cold and windy, and we were anchored somewhere that we wanted to sail away from but it was forecasted to become warmer in the next few days, we would then determine if we could motor the distance we wanted to cover, or wait until the weather warmed up to a safe range for our rigging. Ignoring the small details, like the local temperature relative to where you set your rigging, or the wind conditions and your selected sail plan, will place demands on your rigging in its current condition. The simple switch from a deeply reefed mainsail to a trysail might seem innocuous enough, but this has a drastic effect on the rigging and the stress it will be subjected to. The trysail does not employ the boom, which means the gooseneck will no longer be pushing forward into the mast. This force is normally absorbed by the rigging as the mast stays in column, but when the rigging is slack, this force can demonstrate more of an effect on the mast position. In the case of an accidental jibe, the forces on your rigging are spiked to incredible proportions, and can break well-set rigging. If the rigging is offline because of cold temperatures, the only thing absorbing all of that shock is going to the poor spar! Switching from reefed mainsail to trysail not only takes the boom out of the equation in the case of jibes, but it also spreads the force of the sail over a long length of the spar, similar to how the muscles of your back attach to a long length of your spine.

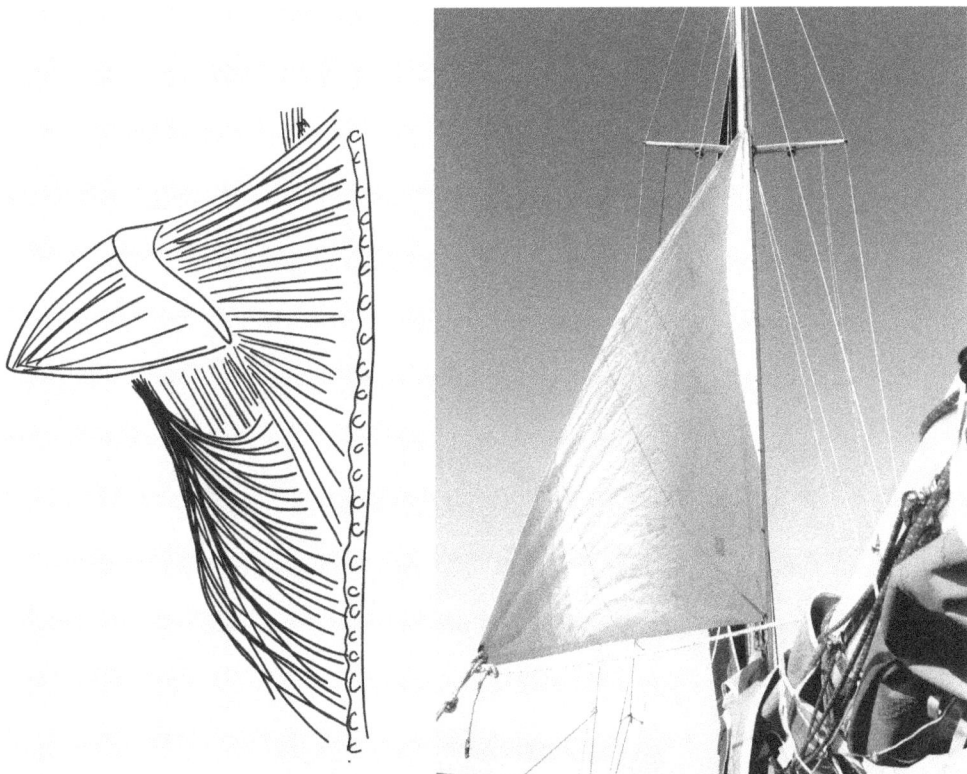

If you have in-mast furling, you might be thinking that you can simply furl in your mainsail until it is at the perfect point, but this is not the case. In-mast furling requires the mast to be absolutely straight and this means that the rig tensions need to be constant. Temperature fluctuations will make synthetic rigging fluctuate in tension as well, which is why this should never be a conversation on what to do with your in-mast furling boat because those boats need to remain rigged with steel.

You might have noticed that there was no mention of linked versus continuous rigs because if you have a linked rig, you can convert it to a continuous rig by replacing the spreader tip or the entire spreader and adding an additional chainplate. It is possible, but it is also a lot of modification and work.

Is synthetic standing rigging right for you? You should now know that answer.

CHAPTER SIXTEEN
A Frayed Knot

Upsizing your steel standing rigging is a big no-no because of the added weight aloft, which makes everything about your sailboat surviving a storm worse. What about your running rigging? Is it safe to change the size of that? Should you replace it with the identical line that your boat came with? We know that the naval architect who designed your boat determined the appropriate size of wire for the standing rigging; isn't it safe to assume that they did the same for the running rigging?

Imagine that you have a ½-inch (12mm) halyard and someone gives you a free ¾-inch (19mm) line. This line is so much bigger than your existing halyard that it simply won't fit through the necessary passages on your boat. With running rigging, you are best to replace with the same diameter of line, that way everything will pull freely as you sail, but you are free to improve the material that the running rigging is made of to improve your sailing characteristics. The gear on your boat is set up to work with a specific size of ropes, and if you go too big, it simply won't fit on your boat.

SHEETS AND HALYARDS

If your sailboat is new or has not had the running rigging replaced, then it probably has mono-colored lines, all black, tan, blue, green, red, and gray. This is because these ropes are the cheapest thing that the sailboat dealer could install on the boat to sell it as a "Sail-Away" setup. This cheap junk is comparable to rubber bands holding your sails in the wind. As the wind builds, the lines will all stretch, and when the puff passes, they will spring back to a shorter length. Proper sail trim is impossible to achieve because the lines are constantly changing length! These lines are referred to as Double Braid Dacron to distinguish them from the good quality lines made of the same material.

Double Braid Dacron doesn't have to be a garbage rope and the good quality ropes will have extra colors woven into the cover to help identify the brand and model. New England Ropes is one such brand that makes a wide range of quality ropes that you can rely on with confidence.

Picture sailing along on a nice day and then a massive gust hits your boat. If you have stretchy running rigging, your sail will turn into a huge bag pulling you to leeward, while if you had infinitely stiff running rigging, the sail would hold its trim and simply carry you faster through the wind. The sail is a wing and the more you can control the shape of it, the more you can control the way your boat sails. It's much more predictable to have a wing that holds its position than one that keeps you guessing.

If you are just getting a feel for things and want to replace your running rigging with the cheapest possible option, you can use Sta-Set for the sheets and Sta-Set X for the halyards. Sta-Set has a braided polyester core while Sta-Set X has parallel strand core. When loaded up, you can experience 1.8% stretch with Sta-Set while Sta-Set X would only stretch 1.3%. The reason not to use Sta-Set X for everything is because it is rather stiff and that makes it uncomfortable in your hands when you are constantly adjusting the sheets.

If you want to upgrade your running rigging but don't want to break the bank, using VPC with its Technora core will decrease your stretch to 0.8%. You just cut your stretch in half with a rope that only costs a tiny bit more! VPC will give you the best bang for your buck, but this is the last time that upgrading doesn't cost a lot.

The rest of the lines we will discuss are all very expensive and provide slightly better performance than VPC. The first one up is T-900, which stretches only 0.5%, then V-100 at 0.4%, then Endura Braid at 0.3%, and, last, Pro PBO with 0.0% stretch.

As you can imagine, the boats in the Volvo Ocean Race are all using PBO rigging because it is amazing and they have the budget to afford it. For the rest of us mere mortals, VPC is the best we should strive for. The price jump from VPC to Endura Braid is a little more than double for the same size line. When your sheets are over a hundred feet long, the price difference really starts to add up. Now, the super high-quality lines are available with solid color covers, but there is no way that the broker installed the most expensive rigging on your new boat by accident. The only way you have an expensive and single color is if you installed it yourself.

CONTROL LINES

Cunningham

While the Cunningham adds a great deal of tension to the luff of the mainsail and thus moves the draft of the mainsail forward, the lines themselves are not under that much load as they are working in the most opportune position to carry out the task. The load is then shared over multiple legs of the same line and is therefore partitioned accordingly. Most Cunninghams will have a 4:1 fiddle block, which means that each individual line is only supporting one-quarter of the total load. For this reason, Sta-Set will suffice for everyone except the most hardcore racer.

Traveler
Positioning the boom athwartship is a critical part of mainsail trimming. You can quickly judge the caliber of a sailor by asking them which sheet trims the main. If they say the "main sheet," then they have simply been on a sailboat before. The quality sailors will say the traveler!

Having the traveler hold your traveler car in the right position might make you envision the mandatory necessity for one of those fancy high-tech, no-stretch lines, but the traveler is a line of many layers of understanding. Just as most sailors think that the mainsheet trims the mainsail, most readers at this point will also be thinking that all lines controlling the sails should be low stretch and VPC at a minimum. Picture a nice controlled jibe where the boom was brought to the center of the boat in a controlled manner, the helm pushed slightly over to catch the breeze on the other side and allow the mainsail to be gently eased out to leeward on a new tack; that was a pleasant situation for the traveler. Now picture reality where your first mate is handing you some nice hot chicken noodle soup that they just made to keep you warm on your watch. It's a cloudless sky with the waxing gibbous moon illuminating the crests of the waves ahead of you as you reach for this little tub of warmth on your cool ocean night watch. Suddenly the wind blows on your other cheek and you see that the waves look to be going in a different direction. You look up to see if the moon is in the same position and the boom comes flying across, as the mainsail snaps into place, occluding your view of the moon. Crash jibes happen and they suck.

If your traveler line were made of a high-tech fiber, it would offer barely any stretch at all and that would mean that the shock load of the crash jibe would have been transferred to other parts of the boat. These shock loads are monumental and they will dissipate into all parts of your rigging, if allowed. Any way that the impact can be prolonged, be it by having the traveler be made of a non-high-tech fiber that has a bit of stretch to it, will add a few milliseconds to the duration of the deceleration. If you paid attention in physics class, then you would remember that this will exponentially reduce the peak force of the collision and therefore reduce the stress on your rig from these little whoopsies that happen.

The traveler is set up with a pulley system, again like the Cunningham, which means that under normal load, the line experiences very little load. The traveler car is the real work horse, the lines simply help hold it in position on the car when the mainsail is being trimmed for upwind sailing. With downwind sailing, the main sheet comes out of the traveler car at such an extreme angle that the car isn't doing much more than acting as a slight turning block to help guide the mainsheet out towards the block on the boom.

During upwind sailing, the traveler car is doing the real work and, thankfully, tacks are easy on the bearings of the little contraption. Downwind sailing is when things can get dicey as crash jibes do occur, despite our best intentions. Since the traveler lines are asked to do very little work when they really matter, having a lower tier fiber doesn't really cause that much of a penalty. The real benefit reveals itself during the inevitable crash jibes that haven't happened yet. With the mainsheet eased as far as it can go, the boom will have plenty of time to build up all the speed it can fathom, and all of this energy needs to be absorbed by something, which you know is going to be your rigging! As long as this peak load doesn't exceed the breaking limit of any individual component, you might escape unscathed, but still shaken up.

The traveler line should be none other than Sta-Set as it actually will hold the load that it is rated to, unlike some bargain brands that claim the rope will hold 1,000 pounds but snap at only a few hundred pounds. You also want to use the smallest rope that fits your traveler system, which is dictated by the small pulleys at the ends of your traveler. These lines will be strong enough to hold the load that will be demanded of them, but also offer some stretch during the inevitable shock loads that will occur.

There are two things you never want to do with the traveler. First, you never want to have it maxed to one side so that it is laying up against the bump stop and the traveler line is not under any load. This takes out all of the protective features you carefully selected for as now the car is up against a rigid part of the track and if something bad happens, something very expensive is going to break as a result. Second, you never want to leave the lazy lines of the traveler slack. When you ease the traveler over, also tighten the lazy side, that way if a jibe occurs, the car doesn't gain a ton of speed as it scoots down the track to its very abrupt end when the lazy line comes into service. Remember that you have a bottom tier stretchy line that is on the thinner side of all the control lines on your boat. Don't push the demands higher by telling it that there will be extra shock loads during the crash jibe. Keeping both lines of the traveler tight and cleated off ensures that when a jibe occurs, only the angle of the mainsheet coming out of the car will invert and everything else about the traveler setup will remain the same.

Reef Clew Lines

These lines should be the exact make and model as your halyard and sheets. If you are using New England Ropes Sta-Set and Sta-Set X for your sheets and halyards, then you will be using New England Ropes Sta-Set for your reefing clew lines. If you opted to upgrade to VPC, then your reef lines will also be VPC. There is no point having an excellent sheet controlling a sail that is being reefed by a bungee cord. The diameter is also important as your reef lines will be controlling your sail in the worst of conditions. It has to pull the smaller version of your mainsail both back and down hard enough that the sail can lay flat, even if the block for the reef line is at the back of the boom. You want to use the biggest line that reef blocks can manage as this will give you the most strength in the line.

Reef Tack Lines
Not all boats will have this in a "rope" version, as they will have a metal hook above the gooseneck of your boom. If the ideal setup is a rigid piece of metal, then you know that any form of rope needs to be just as good as an immovable piece of metal. SK78 Dyneema in the same diameter as your halyard will be your best bet in this application. Single braid SK78 is incredibly strong, chafe resistant, and inexpensive, as compared to other types of lines. The only downside is that the stuff is slippery so knots have to be tied incredibly well or they can pull free on you. If you have a clutch bank, you can add a cover to the Dyneema in the area where the clutch will grab a hold of the surface. Polyester offers a lot more friction than Dyneema does, so splicing the cover over that area will be key to both allowing that section to flow through the clutch with ease as well as ensuring that the cover doesn't remain well held in the clutch while the Dyneema slips out on you.

If your tack line is fixed up at the mast, you can also use a very long soft shackle that will reach through the tack point on your sail and over or through the reef point on your spar. This is all done while the halyard has the luff eased and before the clew line is tightened. Once the tack line is secured, the halyard can be tightened and this will pull it in a vertical direction along with the luff, pulling the mainsail down at the tack. Once the luff is tight, the clew can be tightened. If it is accidentally done in the opposite order, where the clew is tightened before the luff tension is established, then the clew will simply pull the sail aft and you will never get the tack to be close to the mast. This will strain the sail slides and can break them or tear them out of the luff of the mainsail. All of these problems are easily avoidable though by always reefing in a simple fashion: Lower the main to the right amount, securing the tack line, raising the main back up until the luff goes tight, tightening the clew line so that the sail flattens out and the clew rests close to the boom.

As for brands, New England Ropes sells SK78 under the name HTS-78, while Samson sells it under the name AmSteel Blue, and Marlow sells it under the name D12 78.

Outhauls
These lines are very important as they keep the clew of your sail out where you intended it to be. It is also a line that is seldom adjusted on a cruising boat, which

is why the ability to remain where you set it becomes important. You should be figuring out already that the ideal line here will either be a stainless steel braided wire to a pulley system or SK78 Dyneema in place of the stainless steel braided wire. If the setup doesn't have a pulley system, then it probably also doesn't have any wire and is simply a rope that runs directly to the clew car. Once again, this single rope is going to be holding the clew of the sail in its place and therefore needs to be just as strong as a stainless steel wire, and therefore SK78 Dyneema is your ideal fiber for this application. If you don't have a clew car and the clew attaches to the boom with a strop system, it becomes even more crucial that the outhaul be up to the challenge.

Plain SK78 would suffer a hard life as many lines can chafe it as it hangs out in an area where other lines begin to swirl and twirl in high winds. Covering the line with a chafe sleeve or a polyester cover will protect the working part of the line and give you the benefits of making it easier to tie off.

The ideal brands for this will be New England Ropes SK78, Samson AmSteel Blue, and Marlow D12 78.

Downhauls

For once, you get a pass to buy some cheap stuff! The downhaul shouldn't be placed under that much load and therefore any double braid line that is in good condition will suffice. Size is less dependent on duty and more on hand comfort. You are never using this line when things are going well, as the purpose of the downhaul is to pull the sail down when it refuses to be lowered. Try to use something that is as small as possible and is also comfortable for your hands to grab hold of, as this line will also add windage as well as an insignificant amount of weight aloft. If you are debating between 4mm or 5mm, we are splitting hairs as both of these are small lines, so use whichever one feels better in your hands; if you are debating between 6mm and 14mm, you are prioritizing aspects that you shouldn't and you should look at the price per foot of the line so that you can get back on track with the rest of us!

As long as it is rated for marine use, of an affordable price, and comfortable in your hand, then you are good to go. Stick to double braid, though, as this type of line will not twist and get itself fouled as easily on your rigging as a three-lay rope might.

Storm Sail Sheets

This is when you splurge. Fair weather sailors do not have storm sails on their boats because they know better than to go out when it's about to get nasty! For those of us who haven't learned that simple lesson, we have outfitted our boats with all the right gear to handle whatever Mother Nature will blow our way. That gear is expensive, and dedicated storm sail sheets are just another expensive component. These lines should be VPC at a minimum, and they should only be used with your storm sail. They will have their own dedicated sheet blocks that will route them at the perfect sheeting angle and back to the cockpit in a way that avoids chafe on all your other gear so that way you can weather a storm without your sheets failing on you.

It doesn't matter if you rigged your cruising boat with the cheaper stuff or the fact that you will hope to rarely use these sheets—you will pay the big bucks for this extra set of ropes that will live their life coiled out of the sun in a safe dry place so that they are like new every time you go to use them. As soon as they get a little weathered looking, they need to be replaced with new lines but the retired line can happily live out the rest of their days as any other kind of line you need it to be on your boat. It should also be the same size as your regular sheets, that way it meshes well with your existing cockpit gear, be it clutches or winches.

Furling Lines

This is a line that is often overlooked or just plain forgotten about until it snaps and the whole sail comes flying out in a hurry. This line needs to be very strong but also compact and in good condition at all times. If it starts to chafe, the whole line should be replaced as it is not worth the risk. Your massive genoa is never welcome to make an entrance when it is happily furled away for the night. This line should be VPC at a minimum, because roller furling headsails are designed to roller reef, where part of the sail is furled away to reduce sail area as the wind builds. The problem is as the sail is furled, it gets baggy and that makes it hold more air that is trying its hardest to pull the furled part of the sail back out! The furling line is literally all that is holding your headsail in a comfortable setting instead of FULL SAIL. For this reason, it should never be a weak rope. Since size is a limiting factor, as the furler line needs to fit inside the cage on the furler, you are unable to increase the size of the line and simply need to go with a better fiber that can support a greater working load.

Topping Lift

This should be identical to your halyard in size, make, and model so that if your halyard were to fail, you could quickly and easily attach your topping lift to the head of the mainsail and hoist the sail back up to get where you need to go to carry out the necessary repairs. The topping lift in its normal service should never see any of the loads that would require it to be anything more than a plain piece of rope, but having the topping lift as a spare halyard rigged and ready is an example of ways to make your boat prepared for what lies beyond the horizon.

NON-SAILING LINES THAT ARE IMPORTANT FOR YOUR BOAT

Dock Lines

Securing your boat to its berth while you are away is an incredibly important job. If any other part of your rigging fails, it will fail while you are right there to immediately attend to the failure. Dock lines work in the opposite nature, when they are working you are not around, so if they fail, the situation will go unremedied until you or a polite neighbor notices and fixes the problem; by then the boat could have already scraped up against the pier in an un-fendered position and gouged your topsides.

Dock lines need to provide four key points to work properly:

1. They must be strong enough to secure your boat.

2. They must be stretchy enough to avoid shock loads.

3. They must be the correct size for your cleats.

4. They must be small enough to stow.

If your dock lines are too weak, when your boat surges in the slip because a careless powerboater who is oblivious to the damage their wake is causing speeds by, your lines will snap and your boat will no longer be secured in the slip. Best case is that only a few break and your boat stays in the slip, while the worst case is that your boat breaks free and then you get a call from the coast guard informing you that your boat was found on a distant shore and you need to salvage the wreck in a timely manner. With dock lines, strength comes from thickness; the thicker the rope, the higher the breaking strength.

The starting point for choosing the right size of dock line is ⅛ inch of rope diameter for every 9 feet of boat length with the minimum being ⅜ inch because it is still large enough to handle comfortably. You can also reference the table below.

Boat Length	Dock Line Thickness
Under 27 feet	3/8" or 10mm
28–36 feet	1/2" or 12mm
37–45 feet	5/8" or 16mm
46–54 feet	3/4" or 19mm
55–63 feet	7/8" or 22mm
64–72 feet	1" or 25mm

As the dock line gets larger in diameter, it becomes more durable, but also has less stretch to it. If the dock line doesn't stretch enough, then the boat will be jerked around in the slip as the dock lines become taut and snaps the boat. These shock loads will eventually fatigue all the components involved until cracks start to form on the boat around the base of the cleat or, worse, the cleat on the deck or on the dock breaks free.

There are several methods available to provide some elasticity to the dock line system but they are all remedies to one common problem: using the wrong line for the job. Springs, rubber devices, and all sorts of other contraptions exist to add elasticity to the system because the wrong type of line was originally selected. The only type of rope that should be used for dock lines is three-strand rope. It can be made of nylon or polyester, depending on what you want from it, but never polypropylene.

Many times, the wrong type of rope is sold as "Premium Dock Line," which is very misleading. Mega Braid, Mega Plait, and Double Braid lines look prettier and can even be color-matched to your boat's canvas, but these lines offer little stretch.

Double Braid Nylon has a stretch of 6.5%, Mega Braid Nylon has a stretch of 10%, Mega Plait has a stretch of 12%, whereas Three Strand Nylon has a stretch of 16%.

12mm or 1/2 inch rope	Double Braid Nylon	Mega Braid	Mega Plait	Three Strand Polyester	Three Strand Nylon
Strength	8,000 lbs	7,000 lbs	7,500 lbs	7,500 lbs	7,500 lbs
Stretch	6.5%	10%	12%	4%	16%
Type	Double Braid Nylon	12 Strand Nylon	12 Strand Nylon	3 Strand Polyester	3 Strand Nylon

The reason that dock lines are all made of nylon and not polyester is highlighted by the fact that three-strand rope has the most stretch, but polyester has four times less stretch than the same type of rope made in nylon.

When looking only at the nylon ropes, you can see that the way the fibers are oriented allows or prohibits the line from stretching. To achieve the best stretch and therefore the least amount of shock loading, you will need to use Three Strand Nylon rope as it will provide the same strength but with an automatic cushion against shock loads.

Choosing the right size of dock line based on the table above is important, but you also have to make sure that the cleats on your boat are able to support this size of line. If you have a big boat with baby cleats, one of two things happened: either the yacht designer knew the boat was going to be incredibly light and therefore could use smaller dock lines, or the previous owner replaced the cleats with smaller ones for some mysterious reason. Looking at the cleat size you can get a ballpark idea of the correct line diameter. The formula is 1 inch of cleat for every $1/16$ inch of line diameter.

Cleat Length	Line Diameter
3 inch	3/16 inch
4 inch	1/4 inch
5 inch	5/16 inch
6 inch	3/8 inch
8 inch	1/2 inch
10 inch	5/8 inch
12 inch	3/4 inch

Choosing a line that is both the correct size for your deck cleats and for your length will help you avoid getting a dock line that is too thick, and therefore too stiff and too bulky. While the dock lines keep you tied up when you are in port, you also need to be able to stow them when you leave. This means that your lines can't be too thick for your boat and also too long for your boat. Imagine if you have a 30-foot sailboat and you decided to moor it up with 100-foot-long ¾-inch dock lines. Aside from being way too thick for your boat, the absolute bulk of each dock line will fill up your lockers before you have a chance to stow anything you actually need. Dock lines should be a relative size to your boat, that way you have enough to perform maneuvers using your dock lines, but still avoid the need for a spool to coil up all your rope when you leave the pier.

You should carry eight lines on your boat: four lines that will be for the bow and stern, and four lines that will act as spring lines. Bow and stern lines should be ⅔× the length of your boat while the spring lines should be 1× the length of your boat. If you are a perpetual cruiser and always anchoring or tying up in different ports, it would be worthwhile to simply make all eight dock lines one boat length long. This way, you won't run into the issue of trying to figure out which one is the length of your boat for spring lines and which is shorter and belongs on the bow or stern. Any dock line will then go on any deck cleat.

Boat Length	Rope Diameter	Bow & Stern Line Length	Spring Line Length
30 feet	1/2 inch	20 feet	30 feet
37 feet	5/8 inch	25 feet	37 feet
45 feet	5/8 inch	30 feet	45 feet
65 feet	1 inch	44 feet	65 feet

If you are tying up permanently and rarely going to take the boat out, going up to the next size of dock line would be a good idea as stowage is no longer a concern and knowing that you are well tied up is more important. You can also "double up" your dock lines by tying one the normal way and then sistering that line with a second line tied over it on the same cleats but set with a little bit of slack to the second line. The first line will act as the primary line and take all the load in most situations because the second line is a little longer, making it slack when the primary line is tight. When the primary line stretches, the secondary line will come into action and help take some of the load. This is valuable for two reasons: You still have your initial shock absorption, and you also have a spare line in the event that the primary line were to break.

Nylon has one major weakness, which is that it does not like to be wet. Nylon loses between 20–30% of its strength when it gets wet. This is why lines tend to break more in storms that have rain instead of just plain windy situations. Doubling up your dock lines is cheap insurance for these kinds of unfortunate situations.

Anchor Rode

The best material for an anchor rode is not rope, but chain. If you are going to anchor overnight and want to sleep soundly, you need to have an all-chain rode. If you must use rope for some reason, then you should use polyester three-lay rope as it offers more stretch than other types of polyester ropes but will hold up better to being wet than nylon would. Nylon is often sold as "anchor line" but this is the wrong type of material for the job. It will be wet, and that will cause it to lose a lot of its strength and that is far from the ideal situation for the only thing connecting you to your anchor.

Boat Size	Boat Weight	Anchor Rope Diameter
Under 14 feet	Under 1,000 pounds	3/8 inch
14–20 feet	Under 5,000 pounds	1/2 inch
20–30 feet	Under 11,000 pounds	9/16 inch
30–40 feet	Under 20,000 pounds	5/8 inch
40–60 feet	Under 60,000 pounds	3/4 inch
60–90 feet	Under 110,000 pounds	1 inch

If you can carry chain instead of rope, you should use the table below to find the ideal chain size for your boat and then convert your rope rode to an all chain rode, giving you the strength and security of chain and forgoing the concerns of chafe and weakness of rope in the anchor rode.

Boat Size	Boat Weight	BBB Chain	Grade 43 Chain
Under 20 feet	Under 5,000 pounds	1/4 inch	1/4 inch
20–30 feet	Under 11,000 pounds	5/16 inch	1/4 inch
30–35 feet	Under 15,000 pounds	5/16 inch	5/16 inch
35–40 feet	Under 20,000 pounds	3/8 inch	5/16 inch
40–45 feet	Under 40,000 pounds	3/8 inch	3/8 inch
45–50 feet	Under 50,000 pounds	7/16 inch	3/8 inch
50–60 feet	Under 60,000 pounds	7/16 inch	7/16 inch
60–70 feet	Under 70,000 pounds	1/2 inch	7/16 inch
70–90 feet	Under 110,000 pounds	5/8 inch	1/2 inch

Snubber
The snubber is a necessary piece of equipment on a boat, regardless if you have chain or rope rode. The snubber absorbs the shock load from the anchor rode when the boat gets pushed back by wind and waves, which helps protect the anchor from getting pulled out by said shock load. It also takes the stress off the rode as it passes through the bow gear and this one is really important.

With an all-chain rode, the "elasticity" comes from the catenary curve that forms from the weight of the chain. The chain wants to lay on the bottom and rise directly up to the boat, but wind and waves are pushing the boat backwards, which lifts the chain up off the bottom. It hangs suspended in a parabolic form while under load, but as soon as that load lets up, it returns to resting on the bottom. This lifting and lowering of the chain helps make the anchor rode seem to have some elasticity and definitely protects the anchor from any shock loads, as the anchor simply rests in the hole it dug for itself while the chain above waves around. All of this rest comes to an abrupt end if the chain goes taut! Chain has no elasticity to it, so if the catenary curve gets pulled flat, the anchor will be sure to follow.

The snubber gives a bit of stretch to the system so that if the chain goes flat, the rope still offers stretch to the system. Obviously, nylon would be the ideal candidate here as it has the most stretch of all the ropes, but this is an application that will be getting wet. For that reason, polyester three-strand rope is the ideal candidate. It has much less stretch but it will hold up well when it does get wet. Having a longer snubber will allow you to have more total stretch. If the snubber only stretches 4%, that means you will need 12.5 feet of snubber to get just 6 inches of stretch. If you use one of your spring lines as the snubber, then you don't have to store an extra rope on the boat and it is already pretty long, giving you lots of room for elasticity. If you are cruising on a 45-foot sailboat, your spring line will offer almost 2 feet of stretch when loaded.

If you feel the need to carry a dedicated snubber, you should size it to be the same size as your dock lines. It is not going to be under the same harsh demands as the anchor rode itself, as it will not be dragging along the seabed and chafing against rocks. Also, make it at least 30 feet long at a minimum so that you can always let it be very long in severe storms to make life a little easier on your anchor.

Jack Lines
These lines move with you but never move. Jack lines are rigged on the deck to provide you a secure place to clip in when sailing. The idea is that if you fall off the boat, the tether will keep you attached to the boat and aid in your rescue. This is great, as long as you can clip into something! If you have ever clipped in, then you know how incredibly time-consuming it is to clip into something, move a few feet, clip

into something new with the second clip, unclip the first clip, move a few feet and repeat. If you need to get from the cockpit to the bow in a hurry, you don't have time to shimmy down the deck clipping into every little piece of equipment. You need to clip into one thing and run forward. Jack lines attach at the bow and stern, running uninterrupted the whole length of the boat. They should be rigged over sheets and organizers so that you can traverse the deck effortlessly.

They are often sold as flat nylon webbing straps that you tie to padeyes that are installed at the bow and stern, but these have one advantage and a whole slew of disadvantages. The one advantage is that they are flat so they won't roll underfoot if you step on them. Now for the problems: They don't hold up to UV radiation so you can't leave them rigged permanently on the deck. You are supposed to keep them stored someplace out of the sun and if the weather looks bad, run up and rig them. Who wants to rig something at the bow in bad weather?

Jacklines really need to be permanently rigged and always at the ready, which is why they should be made of rope. ½-inch rope makes for a good jackline as it is easy to see and grab when clipping in and it's strong enough to hold your weight if you fall overboard. Three strand polyester is a good choice as it has a little elasticity to it that will help cushion the snap at the end of your fall and it won't look like your other running rigging, which will avoid confusion. How bad would it be to clip into a lazy sheet by accident?!

Tying the jacklines to the bow and stern cleats is an easy and economical alternative to dedicated padeyes. The cleats are already at the extreme ends of the boat and are very well secured. All other lines can simply be tied over the jack lines and that will only further their security. What about the issue of the rope rolling under foot when stepped on? Do you step on your other sheets when walking the deck? Any rope will roll underfoot, and yet you simply avoid stepping on all the other ropes that line the deck of a sailboat. Why not just avoid stepping on this one rope as well? You will get to know your deck very well if you are cruising, well enough that you can navigate your way in the dark moonless sky without stubbing your toe, stepping on a rope, and grabbing onto every handhold you have available to you.

TYPES OF LINES

We have already touched on this topic a little, but now we will go deeper into the different types of ropes that can be used to make your running rigging.

Three-strand rope is made by twisting yarns into bigger strands and then twisting those three strands together. The twisted fibers act like a spring when loaded, unwinding as it gets longer and then recoiling and twisting back into place. This is what gives this type of rope so much elasticity and why it works great as dock lines and snubbers. This elongation before breaking is also what makes this kind of rope dangerous in

high load situations, as the snap back will turn the rope into a whip that can cause serious bodily injury during a failure. For this reason, high load situations should be avoided if another type of rope could be used in that application instead.

On traditional boats, everything was rigged with three-strand rope, and that has given the rope a very nautical feel, but times have moved on and there are better types of rope that should be used instead.

These ropes are easy to knot, hold well when cleated, and are also easy to splice. If an individual strand gets chafed, cut, or damaged, that section of damaged strand can be removed and replaced with a mending splice without needing to replace the entire rope. It is also possible to join two pieces together with a long splice. When carried out correctly, it is impossible to detect and the rope will act as one single long piece.

Single braid rope is made by weaving strands in opposite and equal directions. These ropes are referred to as torque neutral as they do not want to twist in any direction naturally and will not twist when they are pulled tight. Single braid ropes are made out of every fiber imaginable, ranging from cotton to Dyneema, and everything in between. These ropes are easy to splice with specialized tools but individual strands are less easy to repair. If the line gets damaged, the best chance of repair is to splice in a new section where the damaged one once was, or replace the entire line.

The ability to hold a knot is very dependent on the type of fiber. Polyester and nylon will hold knots very well while polypropylene and Dyneema are notorious for having their knots slip out. Splicing these ropes is very quick and easy, as the strands open up to form a hollow core where a fid slips through to carry the spliced part through. The outer part crushes down on the buried part like a Chinese Finger Trap and the harder it pulls, the tighter it holds. With fids on hand, you can splice single braid rope as quickly as you could tie a knot, which makes this type of rope an excellent choice for projects that you want permanent eye splices in the ends of the line.

Double braid ropes usually have a polyester cover but the core can be made of many different materials. The polyester cover makes them much better at holding knots and also makes them easier on the hands when working the line. They are by far the hardest rope to splice. The splices look like magic and then POOF! turn into a beautiful eye splice or an absolute disaster. Their stretch is much less, simply because of the way the strands are woven together, which makes them excellent candidates for the running rigging on your boat. If any damage were to occur to the line, it would be to the cover while the core remains safely protected inside the rope. This is why double braid ropes are the standard on all contemporary yachts. They have the best hand feel, stretch, and strength for the price. They work well everywhere except docking and anchoring, as these are situations where you do want to have stretch in the line.

There are a few steps to take to avoid having any problems in your new running rigging. The first is to use the right type and size of line for the job. The second step is to run things correctly. This means running the lines in a way that chafe is avoided as well as keeping an eye out for changes that could cause chafe to begin developing. Lastly is not using a line for too long. When the time comes to replace a line, ignoring the signs because "it's got a little more time left in it" is a recipe for disaster. You know the line is going, which is why you are telling yourself these sweet little lies to make you feel better while you put your whole boat in jeopardy. Repairing or replacing worn gear will keep everything else working safely, and will avoid you having any epic tales of your own when everything started to break.

CHAPTER SEVENTEEN
Setting Up the Standing Rigging

SHOULD YOU REPLACE YOUR STANDING RIGGING ALOFT OR ON THE GROUND? THIS IS something you need to decide before you start replacing your rigging because the rerig will go very differently depending on which you want to do.

The advantages of replacing the standing rigging aloft are that it is less of a production, there is no need for a crane to unstep and remove the mast. The mast remains in place and the stays are replaced one at a time. This method does take a lot more time as each stay involves its own journey up the mast. If anything slips or falls from the rigger's hand while up there, it can seriously injure someone on the deck and also bounce into the water where it is lost forever. If you are paying someone to do it, their labor time will be much longer, and time is money, so the end price might be about the same. The only difference is that you didn't have to go through the headache of coordinating the haul out, the crane, the land storage for the boat and mast, the wondering if the mast will come unstuck from the step, the worrying as a gentle breeze blows the mast around as it dangles by a single lifting point, or the entire team of people who are involved with this process. A rerig with the mast in place can be done by only a single person. The stay is made and then the stay is taken up the mast to connect, then the stay is connected down on the deck. A rerig done with the mast in place is much calmer and simpler to do, but it has one major drawback: heights.

Everything is harder to do when you are dangling in the air. Your limbs will be clinging to the mast as boats go by and their wake makes the boat sway. A gentle wake for the boat is magnified the farther up you go. Everything needs to be lifted up with you, or you need to have a good helper who can tie it into a messenger line so that you can pull up the one part or tool that you forgot, otherwise, you need to go back down to get it. If you don't like heights, working on the ground is going to be your preferred choice. Even if you like being suspended, you have less control and coordination up high so if the task is very complicated or delicate, it will never look as good as if it was carried out at ground level.

If all you need to do is connect the stay to the tang and then come back down, doing the job aloft will be fine as these are simple tasks to carry out. If you need to install a new fitting and have proper galvanic isolators in place while also torquing the fasteners to a prescribed amount, then you should really be at ground level so that this job can be done with the greatest of care and focus.

CONNECTING STEEL STAYS TO THE MAST

There are three main ways to connect a stay to the mast and as long as you are not changing the method that your mast uses, the rerig will be straightforward and can be done either aloft or on the ground.

The simplest way to connect rigging to the spar is with two plates bolted to the mast and the stay slips between the plates and then is secured in place via a clevis pin. This method provides good articulation for left and right pivoting but does not allow any articulation in the other dimension. Using a toggle at the top of the stay is important as it creates a universal joint allowing the top of the stay to wiggle around in both dimensions, which will avoid stress to develop on the metals and prevent stress cracks from forming. This type of connection is referred to simply as a mast tang.

When connecting the stay to the tang, it is important that the clevis pin head be pointing up when everything is at rest and the cotter pin be on the low side. The goal is that gravity will hold your rigging in place should parts start to fail. Having the head up will hold it in until at the next inspection the problem can be discovered and addressed. Aside from doing what is right, if you assemble the parts so that gravity holds them in place, then gravity will hold it all in place as you get your tools ready to spread the cotter pin's legs. If you have it set up backwards, then you will need to hold it in place to keep it from falling out while you secure it. Working aloft makes it very easy to tell which way is the right way to install it because it will fall out right away if you set it incorrectly, but if you are working on the ground on a mast that is sideways, you will have to think about the orientation on every part you install. You are at ground level so you should have plenty of mental bandwidth to do such thinking as you are not suspended 50 feet in the air looking down at seagulls that are flying below your toes.

On cap shrouds, where the chainplate is vertical and the clevis pin is perfectly horizontal, the head should face outboard so that if the cotter pin were to fail, the windward clevis pin would still be held in place while under load. When you tack and this clevis pin becomes the lazy side, the head will now be facing down and might fall into the water, at which point the cap shroud would start blowing in the wind. How would you like to be notified of a massive rig failure? Leeward cap shroud blowing in the wind or windward cap shroud popping off and the mast collapsing? Always have the clevis pins head up when working so that everything remains under control even as everything appears to fall apart before your eyes. The only time it really doesn't matter is when your chainplate holes are oriented so that the clevis pins run fore-aft. There is no set rule on how these should be oriented but I prefer to set the head of the clevis pin facing forward and the cotter pins on the aft side of the chainplate. The reason is simple, you spend most of your time in the cockpit looking forward and this makes it so that you can quickly peek at the cotter pins from the helm if the thought should ever cross your mind.

If a universal joint at the top of the stay seems like a lot to do to you, it did to somebody else too, which is why T-ball connections were invented. T-ball fittings look like a lollipop that slides into a special fitting on the mast. In the simplest of terms, you put it in sideways and then rotate it as you drop it down to the bottom of the slot. The idea is that the T shape will never be able to slip out while you are sailing so, therefore, it is secure and your mast is well connected.

From previous chapters, you know that the neck of the T-ball is a weak point as the process of creating the ball shape work hardens the metal to the point that it is much more likely to form a crack at the neck. It's not a reason to shy away from this system, but you should know the failure points in your rigging so that you can then inspect it properly. While the T-ball would need to rotate into an impossibly high angle to slip out, this has been known to happen, and to remedy this issue they sell rubber plugs that slip into the slot above the T-ball that keep it in place and avoid having a stay fall onto the deck one frightful day.

If you think there is a better way to connect the rigging in a more secure form, so did someone else. Stemballs are the ultimate "this can't fall apart" setup, but they can also be a little tricky to install. Stemballs use a simple principle to work, the head is bigger than the rest of the stay so the head can't fit down the same hole that the rest of the stay did. Stemball stays need to be installed from the top down, preferably without any lower terminators connected. The wire is pushed down the hole and pulled the entire length of the stay until the head comes to rest in a shallow cup that is specifically prepared for the purpose. Stemball fittings allow the stay to shimmy and scoot in all the directions as the stem head simply slips around in the cup, acting as a universal joint. This system is excellent as there are no parts to fall off of it once it is installed. There can be issues relating to the installation, where bushings or cups might be forgotten and never installed, so do check on your inspections to make sure that all the parts are in there. In very rare occasions, the bushings can crack and break out, and thus not be there in the future, but usually these fittings will last the life of the stay. It is very important that the top bottle screw be fixed to the lower bottle screw in some form because the stay can spin and that would unscrew it from the turnbuckle.

Stemballs are great for linked rigs where the spreader base will have the cup molded into the frame, allowing the compressive force of the cap shroud to be opposed before ever reaching the mast. An advantage of stemballs over T-balls is that stemballs are completely straight, there are no bends at the neck, which means that the load on the fittings is completely in tension and not an inclined beam situation.

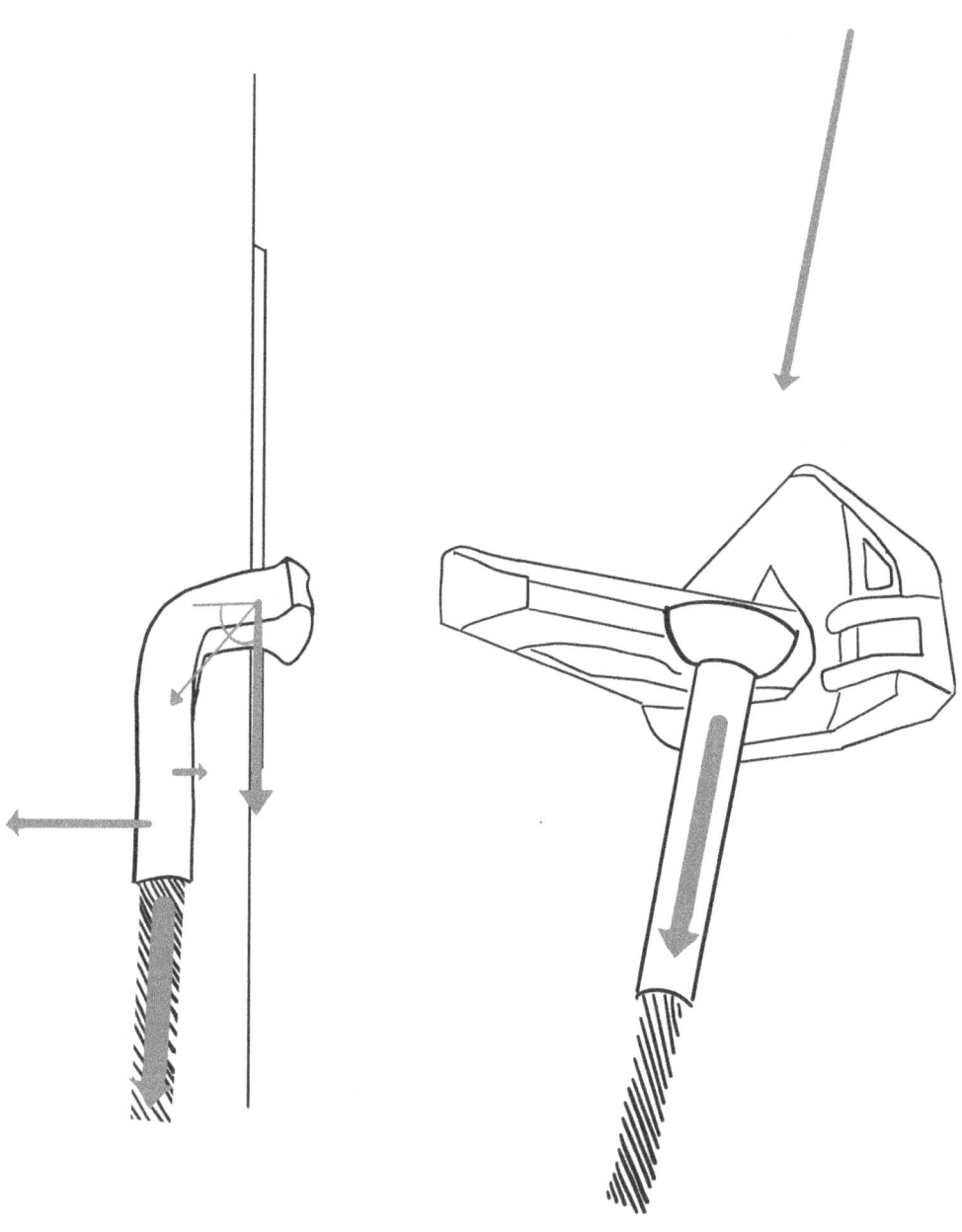

SETTING UP THE STANDING RIGGING

If you have rod rigging, replacing it yourself is probably not going to be a likely scenario. The only way to form the heads on the rod is with specialized equipment that professional riggers who do rod rigging would possess and therefore would also do the whole job. As mentioned before, the dangers of rod rigging are that as the heads are formed, the neck is work hardened and at risk of cracking from the moment it is made. The only reason to stick with rod rigging is if you are racing and the added wind resistance from any other type of rigging would be unacceptable. Otherwise, you should really consider consulting with a local rigger to have your mast tangs converted from rod rigging to any other type. T-ball tangs are easy to retrofit onto a mast, but they run the risk of breaking at the neck because of the same stresses that are created when the head and neck are formed. Stemball fittings are not as easy to convert to. They look like larger versions of the rod tangs, but they use completely different setups due to the increased size. Stemball fittings on the masthead are actually mounted onto the top of the masthead casting and then run down through the top of the mast for a few feet before emerging through a slot on the side. This stretch is impossible to inspect without disassembling the mast! On smaller boats, the masthead casting will have the cap shroud cup just to the side, allowing you to visually inspect this stretch of wire.

The advantage of stemballs is that their components can't fall apart during operation, but the disadvantage is that you can't check on them for early warning signs as you can't see the condition of the metal that is resting inside the cup. The last option that you could convert to if you have rod rigging is the old-fashioned external tang. To carry out this conversion, aluminum pipes would be welded through your mast so that through bolts could be attached to support the external tangs. It is an involved process but the loads of the shrouds are firmly affixed to the mast and distributed over a large area. All the parts are external so inspections are easier, but there are more parts involved that can come apart and disassemble while aloft causing the rig to fail. Which setup you choose to go with is entirely up to you as all setups have their pros and cons, so finding the one that you feel safest with is going to be the best option for you.

Once all of these setups are connected to the mast, the next step will be to connect all the stays to the chainplates. If you have been rerigging in place with the mast up, then you will simply connect the stay at the mast and then come down to that stay's chainplate to connect it down at the deck. If you have been working with your mast on the ground, then once all the stays are connected to the mast, it will be time to lift the stick with a crane and re-step it. Since you are paying for every moment that the crane is present, it would behoove you to label all of the stays so that in the confusion of everything going on all at once, coupled with the roar of the crane engine, you can think clearly and work efficiently, attaching the right stay to the right chainplate without much thought.

For steel rigging, the connection at the chainplate is going to be the moment of truth. If everything was measured properly and, furthermore, cut properly, then the turnbuckles will easily reach the chainplates and can be tightened by hand in moments. The bottle screws should be about halfway through their capacity when the stays are tight. If the turnbuckle is too closed down, then the measurements were a little long, if they are too open, then we are all lucky that the stay even reached the chainplate. Having extra toggles on hand is very helpful at times like these as each toggle can add 1½ inches to the length of the stay, making a short cut long enough to connect everything so that the crane can be on its merry way and stop charging you as you look at the gap between the bottom of your turnbuckle and your chainplate!

If you have swage fittings then the bottom of the stays are completed as the lengths of your stays are already dictated by the cut length prior to putting the end into the swage machine to crush the metal into place. If you are using compression fittings, the wires can be left a few feet longer than needed and then cut in place to the exact length needed. Due to time constraints, deck-stepped rigs that are set up on the ground will typically use swage fittings as this reduces crane time, and that is a significant cost savings. On keel-stepped rigs or rigs that are rerigged in place, you have the luxury of time and can use compression fittings, as they can be cut and set on the boat one at a time to make the rigging the perfect length. Wire is cheap in the grand scheme of things, so having an extra yard or meter per stay will only cost an insignificant fraction of the cost of the new rigging.

Once all the stays are connected and the mast is supported by the rigging once more, the process of tuning begins, which will be discussed further in chapter 19. The short form of it is that you want to get the masthead centered over the boat, so if you are in the water you will use a halyard and a weight to turn your mast into a giant plumb, but if you are on the hard, chances are that your boat is not exactly level so this method would not work. You can, however, use the keel as a visual extension out of your boat on the bottom to delineate what is vertical. Carrying the visual line of your keel up through your boat and into the mast is a great way to get the mast mostly centered until you are splashed and the rigging can be fine-tuned as needed. One last method is to use the halyard and triangles to get it centered. Draw out the halyard so that it reaches the port deck, when you bring it around to the starboard deck, it should touch at the same point. If the halyard is too long or too short on the other side, that means that the mast is tipped away from or towards that side.

SETTING UP THE STANDING RIGGING

CONNECTING SYNTHETIC STAYS TO THE MAST

Synthetic standing rigging always ends in an eye splice. The fibers are too slippery to pinch, grab, or squeeze, so a loop must be made and that loop then needs to pass over something to hold it in place. If you have a traditional wooden mast with knees, then the loop simply needs to be made large enough that it can slip over the spar and engage itself on the appropriate knee. This system is very old, and also very reliable, but you must be sure to install the stays from the lowest stay to the tallest stay, otherwise you will have stays crossing and tangling each other when you go to tension it.

Moving on to contemporary rigging, we will see the same options as steel, but bear in mind that all of these methods were designed for steel rigging and only one of them lends itself well to synthetic. The other methods simply require adapters to make them work.

Mast ast tangs use two plates that capture the end of the stay with a clevis pin. This method is ideal for synthetic rigging because it is a direct swap! The end of a steel stay has an eye in it that the clevis pin goes through, and your synthetic stay also has an eye splice at the end of it. Installing a thimble at the end of your stay will keep the fibers of the Dyneema happy by respecting their need for large radius bends and turns. If you simply put the eye over the clevis pin, it would work for a while, probably long enough that you would think this is a good idea, but it will eventually fail at this point. The sharp bend will stress the fibers and cause them to break. The top of the eye as it passes over the clevis pin would begin to fray and look fuzzy. This would not be caused by chafe, but simply by bending it too tightly. Dyneema has a few simple rules that need to be followed to keep everything happy.

Rule #1: Keep the load to under 15% of the breaking strength to avoid creep issues.

Rule #2: Keep the rope as straight as possible.

Keeping the rope straight is fairly easy to do with standing rigging as the majority of the rope will be in a straight line, running from the mast to the deck. The only two places where this will be violated are at the ends of the stay where the eye splice is made. The eye splice needs to have a few rules followed, or at least attempted to be followed.

The minimum bend of Dyneema is an 8:1 ratio. This means that if your rope is 1 inch in diameter, then your thimble needs to be 8 inches wide. If you are using the metric system and have 1mm Dyneema, then your thimble needs to be 8mm wide. The second part has to do with the length of the eye splice, as this affects the angle of the two parts coming out of the throat. The minimum here is 3:1 but a 5:1 is preferable. This means that if your thimble is 1 inch wide, the length of your eye splice should be at least 3 inches long, but 5 inches would be preferable.

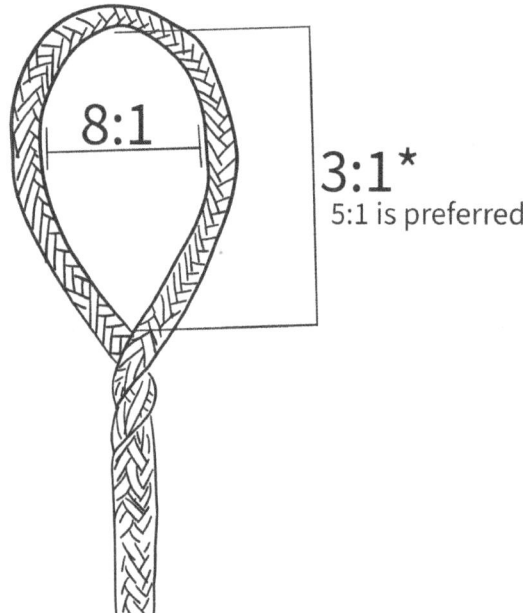

In more practical applications, say your standing rigging needs to be 9mm in diameter. Your thimble will need to be 72mm wide and the length of the eye splice will need to be 27mm long but ideally 45mm long. These are the minimums and you can quickly see that the length of the eye splice is easy to accommodate but the width of the thimble is not. 72mm is 2.8 inches and that is much larger than the largest thimble you will find for 9mm line.

Knowing the ideal and, furthermore, knowing where you deviate from the ideal is important as it will let you know where you need to check to make sure that everything remains copacetic. If you know your eye splices are a little small, then you also know that you need to keep an eye on the outside of the turn for fuzzy fibers. If everything continues to look OK on every inspection, then you are doing fine and can continue sailing peacefully. As you can see, the larger the rope you use for the stays, the lower the percentage of its breaking strength and the lower the creep, but, at the same time, the larger the eye splices need to be. There are structural limitations to how big an eye can be, as it does have to fit against the mast and possibly, next to other eye splices. In these tight installations, going smaller might be your only option to achieve the needs you have impressed upon you.

Lower stays can pose a problem with eye splice sizes, as they tend to be connected to the same tang right next to each other. With steel rigging, this is not an issue as the terminator is very compact, but with a large eye stretching far and wide, this can become a chafe point as the two eyes would rub against each other. The solution here is to use a toggle that rotates the top of the stay 90°. This will rotate the two eyes from being a fore-aft orientation to an up-and-down orientation, allowing the two stays to fit side by side in close proximity and without any interference.

T-balls and stemballs might appear to pose a challenge to connect a synthetic eye splice to a keyed metal hole, but such is not the case. For years, T-balls mast connections have been used for running backstays, which are normally made of rope. There already exists a method of connecting a fiber rope to a T-ball fitting and, therefore, connecting it to the mast.

SETTING UP THE STANDING RIGGING

The ring is too tight of a radius bend for the eye splice in Dyneema, and trying to fit a thimble over the ring can be troublesome, especially once the stay has been loaded and an open thimble would have closed. The best course of action is to use a toggle as an adapter over the T-ball ring, allowing the toggle to connect to both the ring and the eye splice.

When selecting your toggles, be sure to pay close attention to the orientation of the top and bottom of the toggle. Some toggles have the top and bottom in a line, while others offset by 90°. You should use the orientation that aligns with the goals you are trying to accomplish.

Stemballs can be a little tricky to convert as some are open and accessible while others require a bit of creativity. The stemballs that connect outside of the mast are easy enough to connect to, as an adapter that looks like a stemball mounts on the top with a fork on the bottom to accept the eye splice and thimble can be used. Stemballs for the cap shrouds that connect to the top of the masthead casting are much harder to adapt, as the section inside the mast is uninspectable. Using a stemball that connects to a short wire that ends in a fork is preferable as this will allow the section that is uninspectable to be made of wire that should hopefully survive ten years inside the mast, and the eye splice can reside outside the mast.

The reason not to bury the eye splice inside the mast is out of concern that something in there might chafe or break where it cannot be seen. If there is a poorly routed halyard that rubs on the stay, you wouldn't find out until one of the two fell to the deck. Making it all metal in there allows you to receive notice of any rusting that is occurring because the rust would then pour out and down the stay onto the fork after it exits the mast. It also keeps the Dyneema of the cap shroud away from the metal of the slot, in the event that it somehow rubs on the metal and chafes. The windward cap shroud would be rigid and hopefully not in contact with the metal of the gate, but the leeward cap shroud would be slack and could theoretically wiggle around a bit in the breeze, possibly allowing it to rub on the gate and therefore chafe right as it goes into the top of the mast. From the deck, during a quick inspection, you would probably not be able to see this fault, and it would be a very important fault to miss. For all of these reasons, it is best to keep this hidden part made out of metal and have the rest of the exposed stay be made out of Dyneema.

Synthetic standing rigging has a lot more wiggle room when it comes to length. When you are cutting the length of your Dyneema for the stay, all you need to do is some simple arithmetic. First you need to add up all the parts that make your stay longer and then subtract all the parts that will make your stay shorter.

The distance from mast tang to chainplate is your starting distance, and to this you need to add the amount of line needed to make the eye splice and go over the thimble, the amount of line needed for the bury, and then subtract the amount of distance that the toggles, lashings, deadeyes, and turnbuckles will take away.

If you are off by a few inches with steel rigging and turnbuckles, then the turnbuckle will either be closed down too far, or, worse, it won't reach the chainplate because the whole stay is too short. Deadeyes and lashings make this whole process much easier as they grant you a lot of flexibility into the length of the stay, allowing you to add or subtract several feet from your overall length. If the stay is too short, the lashing just needs to be longer to make up the difference and the stay will still reach the chainplate. Building and setting up synthetic standing rigging is much less stressful since you have so much leeway built into the system. It is better to make the stays a little short by accident than a little long because if they are short, you just end up having a longer lashing; too long and the stay may become too blocked with the deadeye and necessitate you to re-splice the eye to make the stay shorter. At the bottom of the system will be the deadeye that connects to a toggle, which uses a conventional clevis pin to attach to the chainplate.

What system you use to tension your synthetic standing rigging is going to depend on your needs, both as a sailor and monetarily. The cleanest-looking installation is going to be just a turnbuckle. This is ideal on headstays where you need the rope to be the longest part so that the sail can attach to the stay as low as possible. It is hard to have hanks slide over a big eye splice and lashings! This also makes headstays the most difficult to fabricate as they must be exactly perfect or the eye splice won't line up with the turnbuckle and be tensioned in the very little amount of distance available by the bottle screws.

Lashings and deadeyes give you a lot of flexibility in the length of the stay. If the stay needs to be longer, all you need to do is have a longer lashing. This changes the range of length available to you for adjustment from a few inches to several feet. As a result, fabricating shrouds becomes an enjoyable process with practically no stress over the measurements as you just need to be in the general area of the measurement to make it work. If you are making a 50-foot-long shroud that will end in a deadeye

With a turnbuckle, you have a very limited amount of adjustment length. A deadeye and lashing affords you greater length of adjustability. This added length allows you compensate for creep that will occur over the years.

with lashings, you will simply make your spliced stay somewhere in the area of 47 feet. This will give you about 2 feet for the lashings and 1 foot for the deadeye. If the stay comes out a few inches longer or shorter, it doesn't matter because you are building in a 24-inch buffer to your system. If you are making this same 50-foot-long shroud and you are using a turnbuckle that measures $15^{7}/_{16}$ inches (38.1cm) when fully open and $11^{7}/_{16}$ inches (27.9cm) when fully closed, this means that you only have 4 inches (10.1cm) of throw in your turnbuckle for adjustment. The spliced stay needs to be exactly $584^{9}/_{16}$ inches (1484.8cm) long when you take it to the boat and install it. You must arrive at this length taking into account constructional shrinkage, constructional stretch, and creep. There are a lot of variables involved in this and if you are outside of that 4-inch (10-cm) window on a 600-inch (1,524-cm) long stay, then you have failed. That is a 0.66% margin of error! This is why making stays that will connect to turnbuckles is very stressful, while making stays that will connect to deadeyes a vacation.

Why use a turnbuckle at all then since it is so stressful when all aspects of sailing are supposed to be relaxing? The reason is turnbuckles are very easy to use to generate the tension needed in your rigging. As you rotate the body, you are causing your rigging to travel along the very long inclined plane of the bottle screw thread. If you need more leverage that is already provided by virtue of this inclined plane, then you can attach a wrench to the turnbuckle body, which will then give you a huge force multiplier. This means that in a matter of moments, you can sit down and turn a fitting that will generate all the tension you need in your rigging. Deadeyes will generate the same tension but instead of using an inclined plane for their mechanical advantage, they use a pulley system. If the loads were light enough that you could simply pull on the lashings and generate enough tension then this would be perfect, but if you have a bigger boat with really high demands for the rig tension, you will need more pull than you can do by hand. The lashings then need to be fed into a system that leads back to the cockpit winches so that you can generate the tension needed, and all this extra rigging takes time to set up and take down. When you are done tensioning, it needs to be tied off with a special knot, known as the shroud frapping knot, to hold the tension in the system while the lashings are disconnected and tied off. This whole setup takes me, the author and inventor of the shroud frapping knot, about 20 minutes per stay to adjust. Compare that to under a minute with a turnbuckle and you can see why turnbuckles are the favorite.

There is a middle ground, though, since these systems do play well together. Having the lashing and deadeye then connect to the top jaw of the turnbuckle will give you the best of both worlds. You can now open the turnbuckle all the way to its longest position and then pull on the lashings by hand to take the slack out. Once the lashings are tied off, the turnbuckle can be turned to tighten the stay and generate the

SETTING UP THE STANDING RIGGING

tension needed. Now, making the stay is easy as you have all the adjustability of the lashings but with the ease of tightening granted by the turnbuckle.

Which system to use depends on your needs, both as a sailor and monetarily. If you plan on sailing in varied temperatures, you will need to adjust your rigging more often as the seasons change. Having turnbuckles would make this much easier as you can simply give the turnbuckle a few turns as the weather cools off and then set it back that same number of turns in anticipation of the warmer days to come. This will allow you to keep your rigging in perfect tune throughout your very long sailing season. The downside of this is cost, as each turnbuckle is a rather expensive piece of equipment and you will need a lot of them.

While converting a Morgan 45, the material cost of all the synthetic standing rigging, all the toggles, clevis pins, cotter pins, lashings, and an entire spare spool of Heat Set SK75 from New England Ropes was $4,400. This was the price for doing the conversion with deadeyes as the boat had 12 stays and therefore would need 12 turnbuckles. The turnbuckles for this boat at the time cost $100 each, and that would have jumped the price from $4,400 to $5,600, a 27% price increase. Using only deadeyes is definitely a less expensive, but more time-consuming method to rerig your sailboat. If cost is your biggest concern as you plan on cruising far and wide on a "beer budget" instead of a "champagne budget," then deadeyes would provide you with a very robust system at a more affordable price. If you can afford turnbuckles, I highly recommend doing the combination of turnbuckles and deadeyes.

CHAPTER EIGHTEEN
Setting Up the Running Rigging

Never forget that the entire purpose of rigging is to hold the sails in the breeze to propel the boat. The spars give you fixed points to hoist the sails and they are supported by the standing rigging, while the parts that move around a lot are supported by the running rigging. The former definition, "Standing rigging is made of metal while running rigging is made of rope," is no longer accurate as now you can have all of your rigging made of rope. While standing rigging can be made of rope, some components of running rigging used to be made of wire. This was a popular setup back in the day, and some older boats that haven't replaced their running rigging will have this for any application where low stretch is desired. If you have any wire components, now is an excellent time to be rid of them. They are known to be difficult to inspect and can break unexpectedly! All the problems that happen to steel standing rigging can occur with wire running rigging, and the recommendation is to never trust the stuff and to replace it with rope running rigging as quickly as possible. In my personal opinion, as the entire world is trying to get away from wire running rigging, if your boat is a simple conversion to synthetic standing rigging I believe that you should heed these warnings on all wire components of your rigging and switch over to rope for everything as it is much easier to predictably inspect the actual condition, allowing you to spot critical failure points long before a failure would surprise you like it tends to do with wire.

By now, you should have figured out what kind of line you want to use, but you still don't know how much of it you need. Line is sold by the foot, meter, or spool, and the more of it you buy, the more you need to pay. Cut lengths can't usually be returned, so it is always better to buy too much than to have a useless line that is too short. Our sample boat will be *Windsong*, a 40-foot cutter that needs everything calculated. The mast height is 45 feet and the bridge clearance is 50 feet.

THE RIGGING HANDBOOK

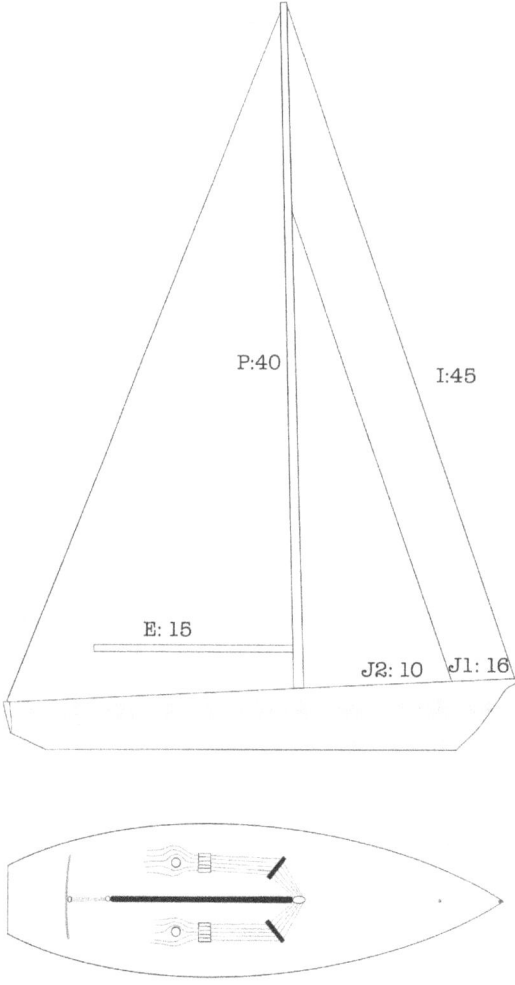

On *Windsong*, the run back to the clutch bank and winches is 14.8 feet so we will call it 15 feet. We will add an extra 10 feet of length so that the line can be easily wrapped on the winch drum and have space to tie the stop knot.

HALYARDS

Halyards, at a minimum, need to be long enough to reach from the gooseneck, up the mast, and back down to the halyard winch. For safety reasons, you should ignore the gooseneck and make it long enough to reach down to the waterline. Should someone fall overboard and need the assistance of a winch to hoist them from the sea, you want to be able to grab any halyard that is available and hoist them out. Imagine if the boat has one "recovery halyard" but the person that falls overboard is the person that knows which one it is and a friend who was on passage with the captain grabs any other halyard and then struggles in vain to make it reach the water to pull the captain to safety! Rope isn't that expensive and the few extra lengths will make your boat that much safer, so measure from the waterline instead of the gooseneck.

To calculate your needed halyard length, you need to know your bridge clearance, and the distance from the mast gate to the halyard winch.

If you have a 1:1 halyard, the bridge clearance is simply multiplied by two, because you need enough to go up the mast and back down the mast. If the winch is on the mast, do not discount the lesser distance of the winch to the waterline, as having a longer halyard during a rescue is a blessing. If your winch is located in the cockpit, trace out the path through all the turning blocks and deck organizers, all the way to the clutch bank and to the winch, then add ten extra feet to this length

SETTING UP THE RUNNING RIGGING

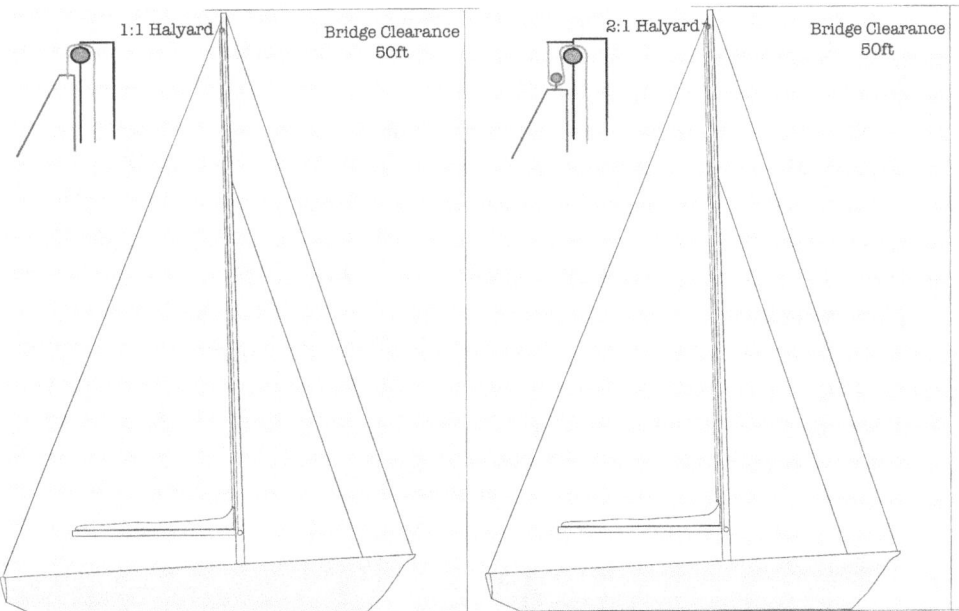

as you will need length to wrap around the winch and for the stop knot. This added length will then be added to the length that is needed for the mast to give you the total halyard length.

If the halyard attaches to the sail with a block instead of a shackle, then you probably have a 2:1 halyard system. These are common on catamarans and sailboats that have very large, and heavy, sails. The 2:1 system allows you to lift half the weight of the sail, but condemns you to pulling twice the amount of rope. Every 2 feet of rope that you pull will raise the sail 1 foot. If you are buying a boat with a tall mast and one of these setups, make sure it has an electric winch for the halyard, otherwise you will be grinding the winch forever to get that sail up into the air. With a 2:1 system, you actually need to multiply the bridge clearance by three as the halyard will start at the top of the mast, run down to the halyard block, run back up to the top of the mast, over the sheave, then back down the mast. This means that your halyard will run the length of your mast three times when the sail is down, and then it still needs to get to the winch. If your winch is at the mast, then you are done with just tripling your bridge clearance, but if your winch is in the cockpit, then you will need to trace its every turn and add up all the lengths until you make the full run to the clutch and then to the winch, then add an extra ten feet of length to let you wrap the winch the minimum of four wraps for light weather or six wraps for heavy weather, and then to tie your stop knot.

THE RIGGING HANDBOOK

1:1 *Halyard Length = Mast Height × 2 + Length to the Cockpit + 10 feet or 3m*
2:1 *Halyard Length = Mast Height × 3 + Length to the Cockpit + 10 feet or 3m*

Windsong with a 1:1 halyard would need to be 50 × 2 + 15 + 10 = 125 *feet long*, while a 2:1 halyard would need to be 50 × 3 + 15 + 10 = 175 *feet long*.

For mainsail and mizzen sail halyards, this method is all you need to do, but for headsail halyards you need to do a small bit of trigonometry. Pythagoras, honored by Freemasons and detested by middle schoolers, created many simple ways to calculate complex mathematical problems that have very practical applications. If you ever asked your math teacher "when am I ever going to use this math?"—today is the day! The headsail halyard runs at an angle from the mast to the bottom of the headstay. Doubling the height of the mast over the water will not do as that does not take into account the forward direction that it needs to travel.

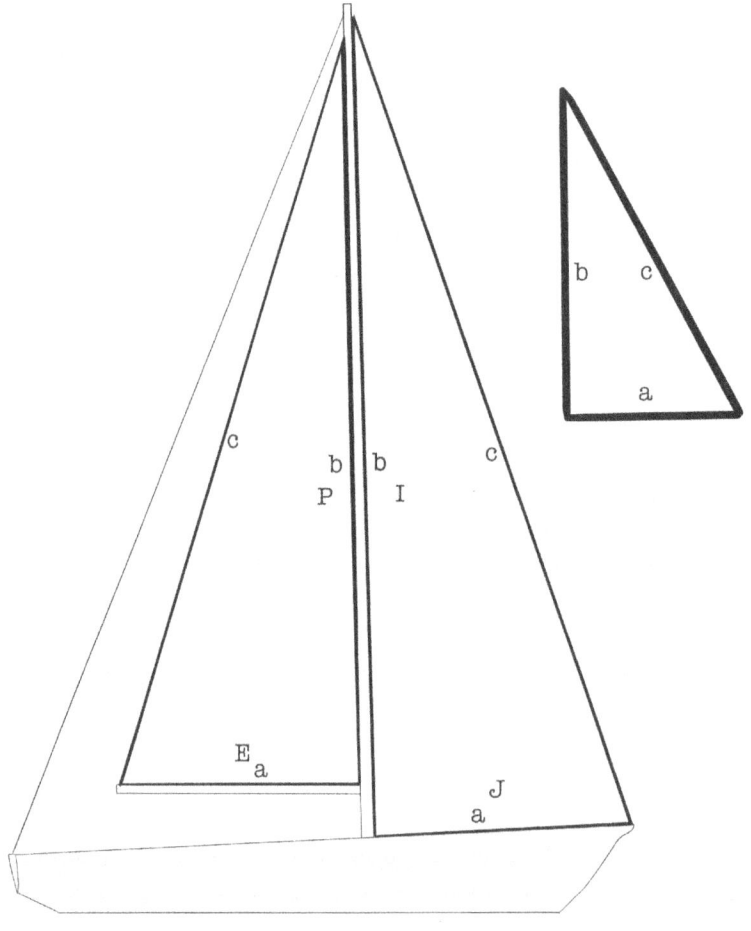

SETTING UP THE RUNNING RIGGING

The headsail halyard sheave height will be *b* (also known as the I measurement), and the foredeck distance (also known as the J measurement) will be *a*, and your halyard length will then be *c*. You know *a* and *b*, but you need to calculate *c*.

$$a^2 + b^2 = c^2$$

$$c = \sqrt{a^2 + b^2}$$

With this simple bit of algebra, you can then calculate the length your halyard needs to run from where the halyard exits the mast, which isn't always at the top of the mast, to the spot on the deck where the head of the sail would rest when lowered. Let's find the headstay halyard length for *Windsong*. The mast is 50 feet above the water, and the boat is 40 feet long with the mast positioned at the 40% mark making it a cutter. This would make the J dimension 16 feet.

$$c = \sqrt{a^2 + b^2}$$

$$c = \sqrt{50^2 + 16^2}$$

$$c = \sqrt{2500 + 256}$$

$$c = \sqrt{2756}$$

$$c = 52.50 \; feet$$

Your jib halyard that runs on the hypotenuse of the triangle comes out to be 52.50 feet, or 53 feet for simplification (always round up, it is better to be too long than too short). Your headsail halyard will need to be 50 feet for the mast section, and 48 feet for the diagonal section, and then added to this will be whatever your freeboard is at this point. The freeboard at the headsail tack is 6 feet on *Windsong*, so you will need add 6 feet to your lengths, making it $50 + 53 + 6 = 109 \; feet$. If your winch is at the mast, then you are done and you simply need to buy the length as calculated, but on *Windsong*, the winch is in the cockpit so we need to add an additional 15 feet to make it to the cockpit and then 10 extra feet for the winch and stop knot. This means that we will need to use $50 + 53 + 6 + 15 + 10 = 134 \; feet$ for the headsail halyard: 50 feet to run up the mast, 53 feet to run the hypotenuse, 6 feet to reach the water from the headsail tack, 15 feet to make the run from the mast to the cockpit, and 10 more feet for the winch and stop knot.

On *Windsong*, the main halyard comes out to be 125 feet, and 134 feet for the jib halyard. Since this boat has its mast at the 40% mark, you should have known that there was going to be a staysail on an inner forestay. This is why the headsail halyards are not calculated from the masthead, but instead from their respective sheave as you would end up with too much leftover on the staysail and just enough on the headsail.

The inner forestay will be represented by I_2 and the distance to the tack will be represented by J_2 on a rig dimension drawing. In our example, the I_2 will be 30 feet and the J_2 will be 10 feet.

$$c = \sqrt{a^2 + b^2}$$

$$c = \sqrt{30^2 + 10^2}$$

$$c = \sqrt{900 + 100}$$

$$c = \sqrt{1000}$$

$$c = 31.62\ feet$$

The diagonal section of our staysail halyard will only be 31.62 feet, or 32 feet for easier math. The freeboard in this section is still 6 feet, and the mast height is 30 feet. The freeboard at the mast is 5 feet, so that all adds up to $5 + 30 + 32 + 6 = 73\ feet$. As you know, if the winch is at the mast, you are done and just need a 73-foot length of rope, but *Windsong* needs that extra length of rope to make it back to the cockpit, so we need $5 + 30 + 32 + 6 + 15 + 10 = 98\ feet$ for the staysail halyard: 5 feet for the freeboard, 30 feet to run up the mast, 32 feet to run the hypotenuse, and 6 feet to reach the water from the tack of the staysail, along with the 15 feet for the run from the mast to the cockpit and then 10 more feet for the winch and stop knot.

TOPPING LIFT

The topping lift should be the same material as the main halyard, but it might need to be a bit longer than the main halyard, as it needs to run to the end of the boom. Doing the math the same way as you did for the jib halyards, you can calculate the length needed for the topping lift, also adding the extra length to reach the water to aid in a rescue.

Windsong has a very long boom, measuring 25 feet and the hypotenuse measures 47.17 feet. The added distance from the end of the boom to the deck is an additional 5 feet, putting it at 52.17 feet. The topping lift length would then be calculated as such: $48 + 50 + 5 + 5 + 15 + 10 = 133\ feet$: 48 feet for the hypotenuse, 50 feet for the mast height to come back down, 5 feet for the topping lift to reach the deck, 5 more feet for the topping lift to reach the water, 15 feet for the topping lift to make it to the cockpit, and 10 feet for the winch and stop knot.

REEF LINES

Reef-line lengths will vary wildly based on the type that they are and how they are run. Single reef lines will be longer lengths than double reef lines, and the shortest are reef lines that are worked on the boom itself.

To calculate reef-line lengths, you need to know the heights of your cringles and their position in the sail. If the reef lines lead to the end of the boom, they will need to be longer than if they have their block on the side of the boom close to the cringle they will be acting upon. On *Windsong*, the sail's P measurement is 40 feet, and the reef points are the traditional 12% of the luff length position. This means that the first reef point is 4.8 feet above the tack, the second reef point is 9.6 feet above the tack, and the third reef point is 14.4 feet above the tack.

The clew will follow suit in the position of the clew cringles, but they will move forward as the leech moves in the direction of the masthead. This added distance will become a factor that needs to be calculated if you have a hollow leech, more so than if you have a massive roach that makes the leech almost vertical.

To calculate the distance, you will simply need to use our old friend's formula to figure out the hypotenuse of the triangle so that you can add in the appropriate amount of line. On *Windsong*, the sail has a hollow leach with no battens in the mainsail, and the reef lines are led all the way aft to the end of the boom.

The boom is 25 feet long and the luff length is 40 feet. If we break it down into a mathematical equation, we get that the hypotenuse is the slope of a line where the rise would be 40 while the run would be 25. Since slope is simply a factor of rise over run, we can calculate that the slope of this line would be approximately 40/25, which is equal to 1.6. This means:

> First reef tack point: 4.8 feet high
> First reef clew cringle: 7.68 feet from the end of the boom
>
> Second reef tack point: 9.6 feet high
> Second reef clew cringle: 15.36 feet from the end of the boom
>
> The third reef tack point: 14.4 feet high
> Third reef clew cringle: 23.04 feet from the end of the boom

As you can see, if you simply assumed that the clew reef lines would be the same as the tack reef lines, you would end up with reef lines that would be too short to work.

On *Windsong*, the single reef line for the first reef needs to run the following course: the boom, up to the first reef clew cringle, down to the end of the boom, down the length of the boom, out the gooseneck and up to the first reef tack cringle, back down to a bail on the mast near the gooseneck, down to the base of the mast to a turning block, along all the deck organizers to reach the cockpit, into the clutch and then onto the winch. This same journey will be repeated for each subsequent reef, the journey is the same but the distances will be greater as the reefs go higher. The length for the third reef would be as follows:

> Boom to Cringle: 23.04 feet
> Cringle to End of Boom: 23.04 feet
> End of Boom to Gooseneck: 25 feet
> Gooseneck to Cringle: 14.4 feet
> Cringle to Gooseneck: 14.4 feet
> Gooseneck to Deck: 5 feet
> Deck to Cockpit: 15 feet
> Additional Cockpit Length: 10 feet
> **Grand Total:** 129.88 feet, or 130 feet of line for the 3rd reef.

When reefing, you will need to crank in all of this line and that is a lot of effort, which is why double reef-line setups are used. The clew reef line would be:

> Boom to Cringle: 23.04 feet
> Cringle to End of Boom: 23.04 feet
> End of boom to Gooseneck: 25 feet
> Gooseneck to Deck: 5 feet
> Deck to Cockpit: 15 feet
> Additional Cockpit Length: 10 feet
> **Grand Total:** 101.08 feet, or 101 feet of line for the 3rd reef clew line.

and

> Gooseneck to Cringle: 14.4 feet
> Cringle to Gooseneck: 14.4 feet
> Gooseneck to Deck: 5 feet
> Deck to Cockpit: 15 feet
> Additional Cockpit Length: 10 feet
> **Grand Total:** 58.8 feet or 59 feet for the 3rd reef tack line.

Having two shorter lines is easier to manage, especially if a jam happens, because you know what part of the sail the jam is in and it can be more easily remedied. You have two lines to manage, but you only have to deal with them one at a time so it becomes a bit easier.

If the reef lines were kept on the boom, then the length would be significantly shorter, as the only line would be the clew line and that would be:

Boom to Cringle: 23.04 feet
Cringle to End of Boom: 23.04 feet
End of Boom to Gooseneck: 25 feet
Grand Total: 71.08 feet, or 71 feet of line for the 3rd reef clew line.

On a side note, if you have an internal pulley system inside your boom that makes your reefing system work, consult your manual to discern how to calculate your exact lengths, as these are too varied to cover all of them. Some have as much as a 4:1 pulley system inside the boom! Your sailmaker will also look at your boom to decide where the reef points can be placed as they don't want to make you a sail that can't be reefed or trimmed correctly. If they mention anything about an internal pulley system, pay close attention and ask lots of questions as they can also help you in determining what setup you have hidden inside your spar as well as where to get more information for the replacement running rigging for this setup.

DOWNHAULS

Downhauls need to be long enough to tie to the head of the sail, route down to the deck with the luff of the sail, and then to a safe location where they can be tied off to keep them from flapping in the breeze; ideally to where the halyard on the winch is located. Having the downhaul lead to the winch for that sail's halyard is crucial to the ease of operating the boat shorthanded. If you can control the release of the halyard and also the rate at which the sail comes down, then you have full control over the sail, no matter the conditions.

If it's blowing well over 40 knots, the pressure on the sail will keep it up in the breeze, where you specifically don't want it as you are trying to reef or simply bring the whole sail down. Releasing the halyard will only make things worse as the wind will catch the rope and pull it out to leeward, possibly fouling in the rigging, which will further the problems at hand. The goal is to release the halyard at the same rate that you are pulling in on the downhaul. Think of it as a circular system, one arm's length of halyard eased and one arm's length of downhaul pulled in. This will keep the halyard in check as the sail comes down, slowly but surely. The downhaul acts on the head of the sail first, as that is where it is tied to, so the top of the sail will be the first part to collapse downwards. At some point, there will be enough flaked sail pushing down, that the weight of the flaked portion will cause the sail to finally come down.

You are pretty much converting the top of the sail into an anvil that is pressing down on the luff, smashing it down out of the breeze. The other advantage of a downhaul in these situations is once the sail is down, and the downhaul is cleated off, there is no risk of the sail riding back up the stay as the wind catches it, because the head is tied down; all you need to do is tie up the body of the sail so that it remains on the deck and out of the wind. With mainsails, having lazy jacks is really helpful in these situations as they will keep the sail on the boom and off your deck while you tie the sail down to the boom.

MAINSHEET

This length of rope is responsible for controlling the position of the boom, and having enough length is imperative to granting the boom full range of motion. To determine the diameter of line you know to reference the size of your blocks, winches, and cleats, but what about the length?

Start by letting your boom swing all the way outboard until it touches your shrouds. This is obviously the farthest that you can ease your mainsail as the standing rigging will not allow the boom to swing any farther away. Now move your traveler, if you have one, to the opposite side of the boat that your boom is situated. If the boom is hard over on the port side, then the traveler car would be positioned as far to starboard as possible. Measuring the distance between the car and the mainsheet blocks on your boom will tell you the maximum distance possible between these two points.

It is important to measure these distances, as boats can have different setups. The mainsheet blocks might be at the end of the boom and therefore need a longer run of rope or they could be midway along the boom and therefore be a shorter distance. The traveler can also be a tiny short track of only a few feet or it can be a very expansive track covering the entire back end of a multihull.

This is the maximum distance that your mainsheet will ever need to cover but it is not the length you want to make your rope to be. If the boom can touch the rigging, then during a crash jibe, the boom can smash into the rigging and break stuff! Having limits built into your running rigging will help you protect the boat so that bad things don't happen during accidents.

Having your mainsheet be just long enough to reach the standing rigging if the traveler car is eased all the way to leeward will be the ideal measurement to work off of. This will limit the absolute range of the boom, but it will also keep the boom within due bounds.

Now that you have your short and long measurements, you want to compare them. If they are pretty close to each other, and you are opting for a high-stretch, low-cost line, know that there is always a good risk that the boom might hit the rigging and buckle. If this is the case, we will address it further when we talk about

SETTING UP THE RUNNING RIGGING

stop knots. If the distances are rather distinct from one another, then you are in a much safer situation should issues arise. Remember that you never want the traveler car to be two blocked against the end of the track. It will always be a little bit away from the end, and these inches that you keep it away from the end will translate into inches that your boom remains away from the standing rigging. Also, the traveler is always eased to the leeward side, so if you have a crash jibe, the traveler will now be on the windward side; so long as you have the lazy side of the traveler taken in and cleated off. A crash jibe will solely mean that the boom will come crashing over but the point that secures it will now be on the windward side of the boat and that will give you even more distance between the boom and the shrouds. The standing rigging will still absorb a shock load transferred to it via the gooseneck, but it won't receive an additional smashing from the boom.

If the lazy traveler line is not cleated, then the traveler car will crash into the other end of the track as the boom flies across. If the car survives the impact, the boom might not if it can reach the shrouds on the other side.

The next step in determining the cut length of rope you will need will be to evaluate the purchase system you have in place on your boat. If the mainsheet leaves the boat and ties to the end of the boom, like some tiny dinghies have, this would be a 1:1 system. If there is a single pulley involved, you now have a multiplier involved and the length of your mainsheet will never be "the distance between the boat and the boom."

Having more blocks adds more mechanical advantage, but at the cost of necessitating you to move more rope to achieve a change in the position of the boom. Where your mainsheet attaches to the boom will also have a big effect on the purchase system you will have. The closer to the end of the boom, the better the leverage, and therefore the lower the purchase system will need to be. You will often see 2:1 systems connected to the end of the boom, but 6:1 systems where the traveler is set forward of the companionway and close enough to be confused with a vang when viewed from a distance.

The distance you have between your traveler car and mainsheet block on the boom will then be multiplied by the purchase system to tell you how much line will be consumed in simply easing the boom to leeward.

Windsong has a 4:1 mainsheet system and the boom is 25 feet long, so we know we will be using a lot of line to ease this boom all the way out. The traveler is only 6 feet wide so that means that the car will be 3 feet off from the midline. If the boom were to swing out 90°, and the boom is 25 feet long, then with Pythagorean Theorem we can postulate that the distance from the midline of the boat to the end of the boom would be about 35 feet long. With a 4:1 purchase, this becomes $35 \times 4 = 140$ *feet* of line just to cover this distance. Being how you pay for every foot of line on your boat, doing a practical test where you swing the boom out to see how far it actually goes and measuring the distance is going to be the gold standard, but if you are far away from the boat and need to buy it before you travel to the boat, it is always better to have too much than too little, so the basic math will be good enough to have you ready to rig your boat upon arrival.

With the distance known, you then need to add enough distance of line for the mainsheet to run through the traveler setup. If there is a winch involved with the mainsheet, be sure to add enough for the wraps that will be needed. To be safe, add 10 extra feet to the end of whatever you calculated that you will need as it is always better to have a smidge more than too little. 10 feet might seem like a lot of waste, but look at it from this point of view: You buy a line that costs $2.19 per foot; 10 extra feet is going to be an extra $21.90. The line you are buying is 162 feet long and will cost $354.78. If you instead buy 172 feet, the price will jump to $376.68, a 6% increase in the price. If your line is too short, it is worthless, and you just wasted over $300, but if the line is a smidge too long, you can either coil up the excess or trim it shorter, either way you can go sailing because the line is long enough to get the job done!

SETTING UP THE RUNNING RIGGING

JIB SHEET

While you might think that the jib sheet needs to be long enough to reach the clew while you are on a broad reach with the sail eased all the way out, you are only half correct. The working sheet needs to reach the clew, but, more importantly, the lazy sheet needs to do the same as well as travel the entire length of the boat. The lazy sheet is the one you need to be thinking about when it comes to the length of the jib sheet.

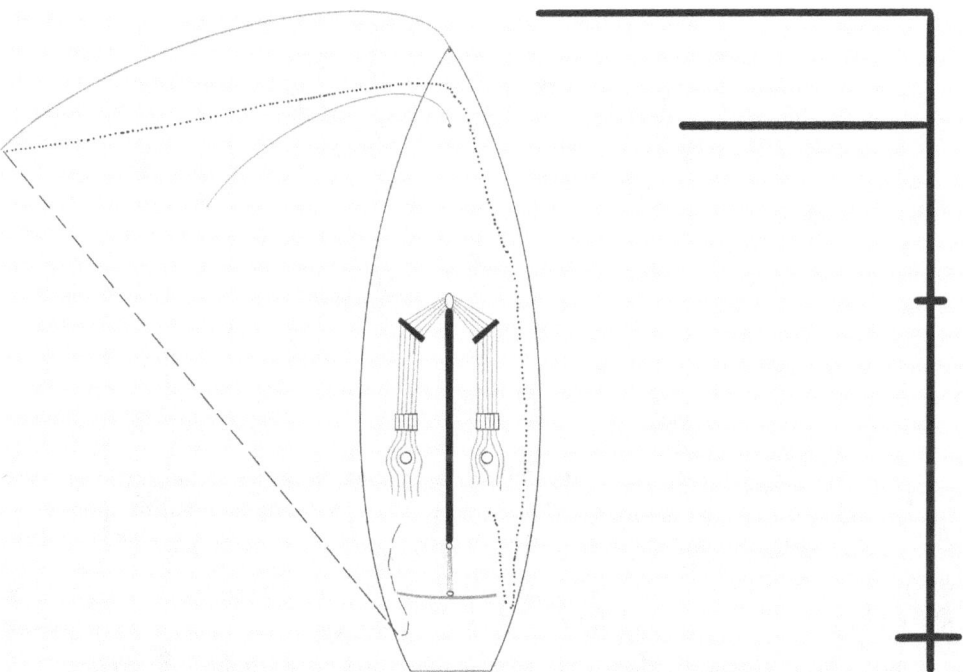

Sailboats are beautiful, but at the end of the day, they all break down into raw numbers on a page. If you ignore all the sleek lines of the topsides, you can better analyze the absolute values of your boat in a way that makes everything make sense.

The first step is to get some graph paper and choose a scale that will work for your application. You want to choose a specific distance that each square will represent so that you can then draw your boat to scale on the paper. If your squares are small enough, you can use one foot per square, but if your paper is small or your squares are big, and you don't feel like getting up to find a page that is better suited for this important task that will save you a lot of money, you can just make each square represent more distance and draw a smaller scale model of your boat on the page.

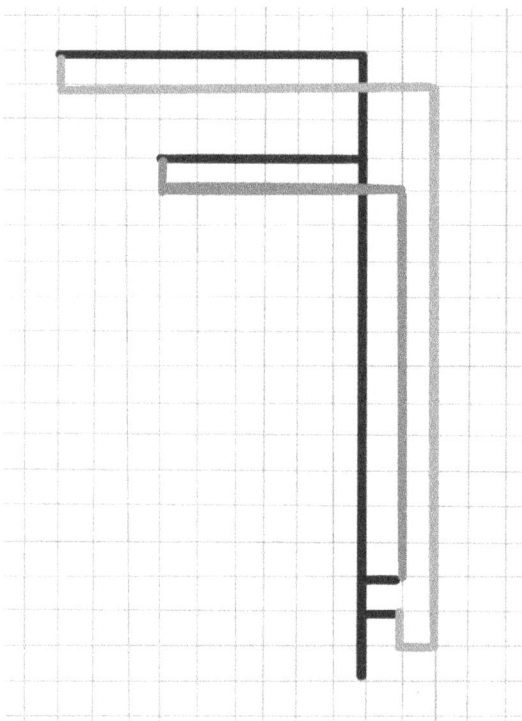

In the example above, the boat is 18 feet long and therefore represented by a vertical line that is 18 squares long on the page. The foot of the jib is 9 feet long while the foot of the staysail is only 6 feet long. This tiny cutter has the jib sheeted to the back of the boat where it returns to a dedicated winch in the cockpit while the staysail sheet block is forward of the winch and therefore the sheet has a direct course to the winch. The jib sheet is represented by the farthest right line on the page, which leaves the winch and then travels aft to the sheet block, then forward to the tack, then laterally to the clew of the sail. The staysail, however, leaves the winch and travels forward to the tack of the staysail and then laterally to the clew of the staysail.

This graphical representation of your boat will ensure that you have enough length of lazy sheet to reach the clew of your sail with the sail eased as much as possible without any risk of being too short. You know that the lazy sheet will never lay parallel to the tack and then turn outboard, but instead will turn as soon as it can, rubbing up against the front of your shrouds or the mast, or if you have a babystay or inner forestay, it will rub against that. The lazy sheet will always make a diagonal run from the last point restricting it to the clew of the sail. The reason it is important to plan it this way is it accounts for all the unanticipated surprises that are sure to show up if you don't use this simple and rudimentary method.

SETTING UP THE RUNNING RIGGING

Picture a huge 180% genoa on *Windsong*, a cutter where the slot between the inner forestay is only 4 feet. On a close reach, the sheet needs to run 6 feet from the cockpit winch to the stern where the genoa block is, then 36 feet forward to get to the front of the staysail, then 25 feet back to run the length of the foot of the 180% genoa with its 29-foot-long foot. Adding all these values up, you will find that the distance run is 67 feet! That is almost double the length of the boat for the lazy sheet, and that doesn't include the length needed to attach the sheet to the sail or for the winch wraps and stop knot. Using the rudimentary drawing of your boat, the winch is located 6 feet ahead of the sheet block, which is at the back of the boat, then 40 feet forward to the tack, and then 29 feet for the length of the foot; this would give you 75 feet of length, which is more than the 67 you would minimally need.

You might be thinking that the magic number to add to this calculated amount is going to be 10 feet, but it's time to mix things up! You want to add 20 feet to the end of your jib sheet because the knot that you should be using to tie the sheet to the sail consumes a lot of line when tied well. This brings our jib sheet length from 75 feet to 95 feet.

The bowline is the preferred knot to use when connecting the sheet to the clew, even though the sheet bend sounds like it should be used for this application. The bowline has some bulky parts to it that can snag on the rigging or the boat, so tying a loose loopy version of the knot will allow the knot to flow over things and get fouled less often. Making the loop of the bowline bigger pays dividends as it then allows the knot to bounce around more when rubbing over the shrouds. If the knot is right on top of the clew, the whole thing is going to get snagged on your forward lower or cap shroud and the tack won't proceed as planned. This big loop greatly increases the amount of line you will need for the knot as every foot of loop is actually 2 feet of line. I like to make the loop about 2 feet long so that the knot is then about 2 feet behind the clew and this means that the loop is going to eat up about 4 feet of line.

The next big waste of line is the tail hanging out of the knot. This portion of the sheet serves absolutely no purpose at all, except it tells you if the knot is not pulling out or everything is fine. The longer it is, within reason, the easier it is to see that everything is fine from the cockpit. If you had to judge the condition of the knot from about 30 to 40 feet, would you want one that has 2–3 feet dangling or one that has 2–3 inches dangling?

Having so much line tied up in the knot means that you will need more line on the other end so that there is enough for handling in the cockpit.

The jib sheets length = the length of the route from the winch to the sheet block + the distance from the sheet block to the tack of the sail + the length of the foot of the sail + 20 feet.

SPINNAKER AND GENNAKER SHEET

Spinnaker sheets will be calculated the same way that jib sheets are calculated, but taking into account that the spinnaker tack can be in different positions depending on the setup that the boat employs. If your tack is at the stem of the boat, then the diagram will look just like the diagram for a genoa. If your tack is set on a pole or bowsprit, forward of the bow of the boat, then your vertical line in your drawing will need to be extended the appropriate number of squares. You will hear people state a rule of thumb for these sheets, stating 2–2.5× the boat length but the truth is that it really depends on the application. If you have a mizzen spinnaker, the sheet will not be anywhere near as long as the sheet for a spinnaker set at the end of a 5-foot bowsprit.

Do the drawing and calculate the actual value, but since the lazy sheet will be routed around the outside of the headstay, give yourself a little extra wiggle room with your calculations. Instead of adding 20 feet to the total length, you will want to add 30 feet. Spinnaker and gennaker sheets are smaller than your normal sheets to be lighter, and this means that they will not be as expensive per length even though they are made out of really high-performance fibers.

DRIFTER SHEET

The drifter sheet needs to be calculated the same way as the genoa sheet was, plotting the lines on the graph paper and calculating everything to a tee. Unlike the spinnaker and gennaker sheets, these sheets can be made out of the same type of line that you are using for your regular sheets, but they can be a smaller size, so long as it is still comfortable in your hand and still works in your self-tailing winches.

Spinnaker and gennaker sails are used by people who want to go fast and squeak out every last bit of performance from their boat. Drifter sails are used by people who just want to keep sailing when the wind dies down. These sails are not used in high winds, and they are not used for high speed. The cloth stretches a lot so there is no point in using low stretch sheets. Plain Sta-Set will do the job just fine and can help you distinguish the drifter sheet from the jib sheet in the cockpit spaghetti that is inevitably going to form.

Drifter sails are massive and have a huge foot, so be sure to account for that when you draw up your diagram to determine the length for your sheets. These sheets only need to have an additional 20 feet added onto the calculated distance as drifters are pretty much just genoa sails made out of spinnaker cloth.

STAYSAIL SHEET

These sheets will be considerably shorter than your jib or genoa sheets because the sail is closer to the cockpit and therefore the lazy sheet has a shorter run to make. If your staysail uses the traditional sheeting method, then you will calculate the length the same way as you did with the jib sheet; but if the staysail has a self-tacking setup, then you will need to factor in the purchase system of the self-tacker and calculate it more along the lines of what you did with the mainsheet calculations.

Self-tacking systems vary wildly, some use a traveler setup with a car, others simply have multiple blocks on the deck, others just make a bridle that the sail slips through. Whatever system you have, you will want to break it down to the most basic principles and calculate it from there.

Always start with the drawing, marking the positions of the blocks on your deck and sail in the diagram, true to the scale. This is where making each square equal to 1 foot really helps as it makes it easier to translate the deck onto a sheet of paper. If one of the blocks is not located exactly on a 1-foot interval, make your life easier by fudging the distance to the next foot. Always err on the bigger side, that way you don't end up short at the bitter end. If the distance between the blocks is 4.5 feet, draw it as 5 feet.

Once the blocks are all demarcated, you can then position the sail arbitrarily to the port side and place the clew the length of the foot away from the tack point. Since these lines will be part of a purchase system, you don't want to make the same 90° lines you were making before because it will overcompensate too greatly. Draw your lines diagonally and use Pythagorean Theorem to find the distance that lines will run as the lines will be running along the hypotenuse of the right triangle that will be made by the block and the sail. The sail will be b while the distance aft to the block will be a and the sheet length to the clew will be c. The sail can extend laterally the entire width of the deck if it makes it easier for you to visualize the right triangle, and since each square is 1 foot, finding the lengths of the sides of the triangle will be as easy as counting.

When you have finished your math for the self-tacking portion of the staysail sheet, you will then need to add in the run back to the cockpit winch and add an additional 10 feet to the calculated length.

TRYSAIL SHEET

The trysail is the most important sail your sail boat will have, as this is the only sail that will save your personal transom in a mighty storm. To show this sail its due respect, it should have its own properly sized sheets that are long enough to do the job at hand but short enough to not be in the way during a storm.

The same principle applies, even though this sail will never be flown directly to the side. It will always have a bit of a curve to it and will usually be sheeted in close to the boat. The important thing is for the lazy sheet to be long enough to reach the clew without adding any additional stress to the situation.

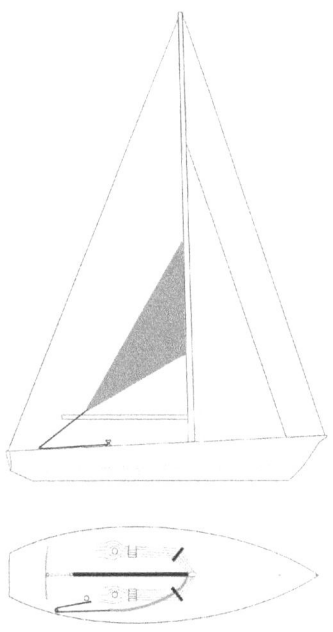

The line drawing will be the same as you did for the jib sheet, but remembering that the sheet block position may be well aft on the boat. The drawing should have the line leaving the winch to the sheet block, then traveling to the mast, and then out laterally to the clew. If the sheet is long enough to do this, it will be long enough to manage well in a storm without cutting across your bimini and dodger. Adding 10 extra feet is a good plan as the sheet will be wrapped extra times around the winch to give it better grip on the line during the storm as it blows over. This extra distance will be greatly appreciated in these trying situations. Remember, the minimum number of wraps on a winch is four, but during a storm you want to have at least six wraps that way the drum is taking the full load and not the self-tailing mechanism of the winch.

CUNNINGHAM

The Cunningham is a small purchase system that has a short distance to cover and provides a lot of adjustability to the position of the draft of the mainsail. They involve a pulley system that will dictate the multiplier you need to consider when calculating the length needed. Most Cunninghams use a 4:1 purchase system, so that means that if you have 4 feet between the bottom fiddle block and the Cunningham cringle, you will need $4 \times 4 = 16\ feet$ of line just for this part. You will then need to evaluate where the line will run to be operated. Does it stay at the mast or does it run back into the cockpit? If it stays at the mast, then adding 5 extra feet to the system will be plenty, as a long tail will become a liability instead of an asset. If the line leads back into the cockpit, then you need to measure the length of the entire run and then add 10 extra feet so that you have enough line to stow in a bag. There is nothing worse than a very short line coming off the clutch bank that just sticks out there because there isn't enough to stuff it into the sheet bag under the winch!

VANG

The vang should be calculated the same way as the Cunningham, measuring the distance between the two blocks and multiplying it by the force multiplier of the pulley system. As with the Cunningham, if it stays at the mast, only add 5 feet to the length, but if it runs back to the cockpit, add 10 feet to the calculated length.

SETTING UP THE RUNNING RIGGING

TRAVELER LINE

This is an easy line to measure and calculate as its run is completely prescribed. Multiplying the length that the traveler car can travel by the pulley system will then tell you how long you need your line to be. As before, adding the beam of your cockpit + 3 feet is handy as it will give you something to grab onto when you need to pull the traveler to the other side.

Say your traveler track is 10 feet long, has a 4:1 system, and your cockpit is 10 feet wide, the length of the line would be as follows: $10 \times 4 + 10 + 3 = 53$ *feet*. This would give you the 40 feet for the traveler system, plus the 10 feet to allow you to work the windward side of the traveler from the leeward side of the cockpit, and 3 extra feet for the stop knot and to have a little extra to grab ahold of.

If you are not sure what force multiplier your pulley system has, there is an easy way to calculate it that involves barely any thinking: count the number of lines that run between the two pulleys. A 2:1 will have two lines between the pulleys, a 3:1 will have three lines, and a 4:1 will have four lines. You don't have to think about the pulley arrangement or how the rope is routed, just count the number of ropes you see stretched out before you and that is your number.

FURLING LINE

Furlers are very popular on boats, they take all the work out of working a sail and allow you to control everything from the comfort of the cockpit. These lines have the most important job on the boat, as they keep you out of trouble. Think about it, if any line breaks, what happens to the sail? Normally it falls away and out of the wind. If a halyard breaks, the sail falls to the deck; if a sheet breaks, the sail turns into a flag; but if a furler breaks, that sail is bursting onto the scene like the Kool-Aid man!

Having a larger sail with a longer foot means that the furler needs to rotate more times to furl it away completely and that will require a longer furling line to make it spin more times. The number of rotations is dictated by the length of your furling line, the longer the line the more times it will make the furler drum spin around.

In high wind situations, the sail will be loaded to extreme levels as you furl the sail away and this means that it will wrap itself tightly onto the furler. This also means that it will take even more wraps to get it to furl away completely. Even when you get the sail furled all the way in, you want to go a few more turns to make the sheets wrap around the furled sail to help hold it in place and prevent it from unfurling accidentally if the wind catches it at just the right angle. As you can see, the correct answer to "how long should your furling line be?" is actually "how much furling line can you fit on your furler?" You will never be sad that you were able to give the furler one more twist for good measure.

There is one limiting factor that will determine how long the furling line can be, and that is the size of the cage around the furler drum. Using a thinner, high-quality line that has the same strength in a smaller package will allow you to have a longer furling line. Looking at the ropes for sale at your local chandlery, you can see how the bigger the rope, the bigger the spool. If you use a big strong rope, it will take up too much space, while using a smaller rope will allow you to fit more line onto the drum.

Start by looking up the specifications for your particular furler and see the minimum line strength that the manufacturer recommends for their equipment. Then look at better rope choices that use high-tech fibers such as Dyneema cores to see how much smaller you can go with the same breaking strength rating. Next comes the guessing game where you buy some cheap garbage brand line in a massive length, way longer than you ever think you would need, but in the diameter of the new line you are going to buy; and you start wrapping it around the furler. Make the wraps nice and neat so that you can fit as much line onto the drum as possible. When the drum is full, run the line back to the cockpit and around a winch drum to see how much line you actually need to have. Since you have an actual line in front of you, add a comfortable tail to the line so that you have plenty to work with when you first load it on the winch. You don't want to have the bitter end sticking out of the clam shells of the self-tailing winches when you have to furl your sail in! Give yourself about 10 feet past the winch as this will be a comfortable amount to have on hand, or if someone farther from the winch wants to help giving the line a pull.

Once you have your length figured out, you can take this sample piece to the local chandlery and measure it out, purchasing that same length of the good stuff. This will give you the most turns possible for your furler and takes all the guesswork out of calculating it.

Why can't you accurately calculate the furling line? Because as the drum fills with rope, the diameter is increasing and this increases the circumference. The rope will lay flat as it gets compressed so each subsequent layer isn't really increased by twice the diameter of the rope. Calculating it is possible but it is just complicated as there are too many variables involved when you could just as easily do a proper test and buy the right amount of the good line one time.

Since you are going to be packing the furling drum as tight as it can be, it is imperative that you be sensible about unfurling the sail. All too often, you will see sailors release the furling line and start pulling on the sheets. As the wind catches the sail, the whole thing shoots out like a magic trick. Everyone sees the sail popping out but no one notices that the poor furling line wrapped up fast and loose. When that sailor tries to furl the sail in, the line will start to pull tight against the loose wraps of line in the furler and this will cause the line to burrow down under other wraps, which will eventually in the process start to tighten down over it. This is an override and now the furler is jammed with the sail stuck at whatever point it made it to. This entire fiasco could have been avoided by simply maintaining pressure on the furling line as the sail came out, making it wrap tightly around the drum and without any chance of creating an override. The easiest and safest way to do this is to wrap the furling line at least three times around a winch and hold onto it as you go pulling out on the sheets. As the wind catches the sail, you can then stop pulling on the sheets and focus just on the furling line if you are sailing single-handed. If you have a crew, someone else can tend to the sheets while you focus on slowly easing the furling line over the winch and letting the sail unfurl in a controlled and steady manner. This will also give you very tight wraps, which will reduce the bulk of the furling line, a very important thing to do when you are planning on stuffing the drum as full of line as it can manage. Furlers are wonderful when they work well; and most of the times that they don't work well, it's due to user error.

DOCK LINES

Dock lines should be sized according to the length of your boat. While the breaking strength of each line will be far inferior to the weight of your boat, it is important to remember that no individual dock line will be holding your boat suspended in space. The boat is floating and the purpose of the dock lines is to hold the boat in a relatively static area of the water. They should never be left to work alone as there should always be a few lines holding your boat and therefore sharing the load. In storms, where winds are high and waves can cause the boat to surge around in the slip, adding extra dock lines or doubling up the individual dock lines will help tremendously.

If you plan on doing some light cruising and visiting other marinas in the area, having dedicated bow and stern lines that are two-thirds the length of the boat and then a set of spring lines the length of the boat will allow you to comfortably tie up in a variety of slips in your area. The trouble is keeping the short lines differentiated from the long lines and making it a system that anyone who doesn't know your boat can do. Imagine you have family or friends on the boat for a fun cruise and when returning to port, Uncle Jim wants to help get the lines ready. He puts a line on each bow and stern cleat and two lines on each midship cleat, you are ready to tie up! When you get to your slip, Mark walks over to catch a line and help you tie up but suddenly the lines are not long enough to reach the cleats. Uncle Jim put the long lines on the bow and stern and the short lines on the midship cleats. Now you need to swap out the long lines for the short lines and everything becomes a lot more work.

If you plan on cruising in varied areas or will be tying up a lot, having eight lines that are the length of your boat is a great plan. You will be tying up a lot and you will have a lot of lines on your boat. Trying to keep the long and the short lines identified or separated will become a constant annoyance when you could just as easily have eight lines that are all the same. This will also give you some added benefits, as the bow and stern lines will now be longer so when you find yourself docking in challenging situations, it becomes a little less awful. A longer line on the bow and stern means that you can toss a line from farther away and have a better chance of someone on shore catching it, as well as the ability to walk a line down the length of the boat.

Imagine trying to tie up on the leeward side of a wall in a massive blow. You get the bow line to someone on shore and they pull the boat close to the wall. If the stern line were long enough, they could actually take the stern line from the bow and walk along the wall to pull the stern in. Midship lines will also help in these moments as the combined efforts can bring you safely to berth without any use of the motor once the bow has been tied up. If the stern line were two-thirds the length of the boat, this plan would never work as the stern line would be just too short.

In areas with massive tides, it is also best to have longer lines as you will rise and fall a great deal when tied to a fixed pier or sea wall. Crossing your lines and running them diagonally will help hold you in place as the tide changes without the risk of your boat bumping into the pier or turning your dock lines into guitar strings as the tide rises and falls. If you are in these areas, look at what the local fishermen are doing and emulate their added lengths.

SNUBBER

If you are going to be doing some extensive cruising, you will also be doing some extensive anchoring. Having eight identical dock lines will pay dividends when anchoring for a storm. The dock lines can also become your snubber line by tying a magnus hitch to the chain and the other end to the bow cleat. The snubber will give you the elasticity your anchor desires in your all chain ground tackle and also help prevent the chain from accidentally running out should the chain stopper or windlass gypsy fail.

In light weather, the snubber can be tied short so that the knot never even touches the water, but in heavy weather the snubber can be let out to its full length. If you need to let out more chain because conditions are deteriorating or you didn't have enough chain out in the first place, you won't need to waste time and effort bringing the chain in so that you can disconnect the snubber only to move it a few feet on the chain. Instead, you can release the snubber from your bow cleat and let it remain tied to the chain for the rest of the storm. If you tied a good knot, it will still be there after the storm passes and can be retrieved when the weather improves. When you let out the amount of chain that you now want, you can then take another dock line and tie it up in the same way and fashion as you did the first one. If the storm is particularly nasty, you can even double up the system with two snubbers, one tied to each bow cleat. This setup will let you extend your anchor chain somewhere between four and eight times during the storm, depending on if you are using a single or double snubber.

Dock lines are absolutely worthless when you are away from the pier and if you are cruising, they need to be able to serve other purposes as well. Acting as a snubber is a great job for them.

PREVENTER

Another wonderful task for the dock lines while underway is to serve as your preventer. The preventer should be tied to the mainsheet bails on the boom as this is the place where the boom is known to be strong enough to take a massive load and oppose the force of the sail. Think about it, the sail is always pulling away from the mainsheet and the boom doesn't buckle in the process. If the mainsail were to become backed, a preventer in this same attachment point would simply act as a mainsheet and hold the boom against the desires of the mainsail.

The preventer line should be routed forward from the boom at an angle that is effective to resist the force of a backed mainsail, and then tied off to a very strong cleat. Ideally, the bow cleat would serve best but if angles allow and distances prohibit, the midship cleat can serve this purpose well.

Two main situations where preventers become incredibly useful while the mainsail is lowered are to pole the boom off to the side while at anchor to avoid shading the solar panels and to get the boom with all of its associated rigging out of the way of the trysail during a storm.

Poling out the boom is straightforward and doesn't need much explanation but the trysail scenario may take a moment to sink in. The trysail should have its own independent track that runs up the mast next to the mainsail track. This track is either to port or starboard of the mainsail track because the mainsail track is installed in the middle. If the track is to starboard, that means that on the port side of the trysail will be a stacked mainsail, and all the gear and equipment that is run near and around the gooseneck. If the boom is centerlined, the starboard trysail sheet will simply run straight from the sheet block to the trysail clew. The port trysail sheet needs to be fished forward of the topping lift, aft of the last lazy jack, and over the boom. This will allow the trysail to flop from side to side without fouling on the topping lift, and hopefully not fouling the lazy jacks that always manage to be in the way during these situations. The lazy jacks can be lowered out of the way to make room for the trysail, but now you have additional work to do as the winds build.

A simpler approach is to take everything that's in the middle and put it over there! Polling the boom far to port will move the boom, mainsail, topping lift, and lazy jacks out of the way and make a clear area for the trysail to be hoisted and to fly in. As with everything, there are limits to how far you want to take things. If you pole the boom out too far, it will dip into the drink, while if you don't pole it out far enough, it will still be in the way. You will have to play around and find the right position on your boat and with the aid of the dock line as a preventer, you can position the boom in this position and hold it in place between the preventer and the mainsheet.

STOP KNOTS

Stop knots serve the simple purpose of stopping the line from slipping out through the blocks and going about their merry way. Most people place stop knots at the very end of a line, that way the line can't pull away. This is a very common and incorrect setup on jib sheets. Tell me, if the wind was strong enough to pull the line free from your hands, what makes you think you are going to do anything with a chunky knot sticking out from a block on the side of the deck?

Stop knots should be placed in the line far enough forward that there is enough line after the stop knot to comfortably wrap the line around the winch and pull it back into normal control. Yes, the stop knot will be annoying as it passes through the winch, and you know you will have to take the line out of the self-tailing mechanism as there is no way that the stop knot will make it through the clam shells. Having a long enough line that the normal working lengths never involve the stop knot will be critical for your comfort. This is part of why determining the longest length necessary for your sheets and then adding additional feet to it will help you avoid encountering a situation where the stop knot will become a hindrance instead of an asset.

The stop knot on the mainsheet should be positioned at such a length where the stop knot reaches the traveler car or first mainsheet block on the boat right before the boom contacts the shrouds. This is going to prevent catastrophe and make a bad situation just a warning as to how much worse things could have gone had you not been set up properly.

Stop knots are your limiters to keep everything on your boat safe. Performance cars have rev limiters to keep the engines from spinning too fast and blowing up, and your sailboat will have the same thing on each and every sheet and halyard to prevent a line from slipping away and creating a situation.

Always set the stop knot far forward of the bitter end, with enough tail behind it to get at least three wraps around the winch drum. This will let you get the sheet back on the winch and crank it back into submission.

Several types of stop knots exist and the most common is the Figure Eight knot. This knot is easy to tie but doesn't lay in line with the rest of the rope. It also offers considerably less bulk than other stop knots, which means that it could possibly be pulled through a large enough hole with enough force applied.

The double or triple fisherman's knot will provide a much better stop knot and it will also look much prettier in the process. The line will have a nice barrel on it

SETTING UP THE RUNNING RIGGING

at key locations to prevent issues and keep our rigging in check. Double and triple fisherman's knots will have the line entering and exiting the knot in the very middle and simply add a lot of bulk evenly around the line.

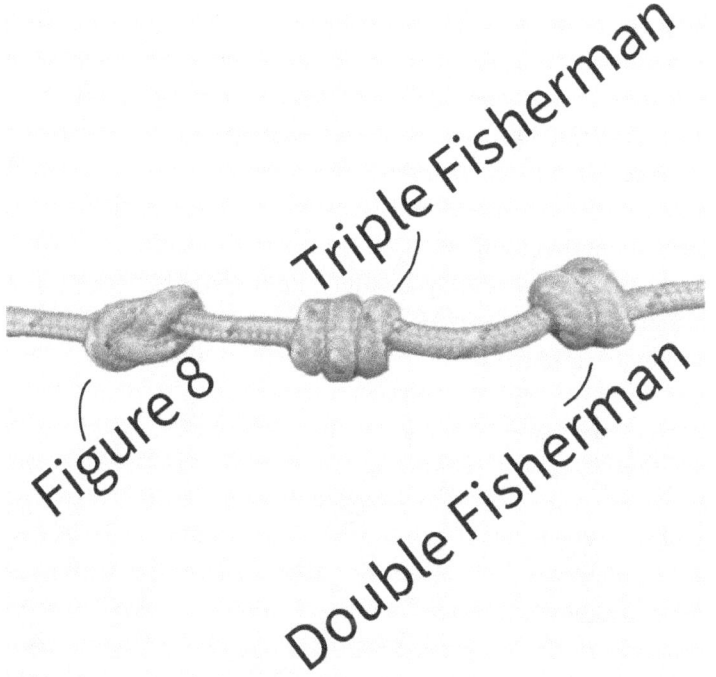

BUYING THE ROPE

After all your calculations, the time will finally come to lay down your hard-earned cash and trade in a stack of money for a bunch or rope. Ironically, rope gets less expensive as the price climbs. When you see rope listed, you will see the price per foot. You might also see some bulk discounts shown when you buy over 100, 300, or 600 feet of rope. 100-foot discounts are a courtesy the chandler is granting you since they know you are about to pay a lot of coin for this purchase. The 300-foot and 600-foot discounts are because you are buying a full spool of the rope and the chandler is glad to move that much inventory without running the risk of getting stuck with a short, unsellable length of rope.

When you buy rope in these long lengths, you will save a considerable amount of money, but you will also have a lot of the same looking rope. Some sailors like to have each rope be a unique color to help identify which rope is which when it forms the inevitable spaghetti that you will cook up in your cockpit while having the time of your life, but others choose to save some money and buy huge amounts in the same color to take advantage of this discounted approach.

If you decide to go with a single color, know that each line of a different size will start the count over so you can make all your 9mm lines red, your 10mm lines green, and your 11mm lines blue. This will give you a little clarity in your situation as you are working the sails.

When buying the rope, it is also better to order the cut lengths instead of the grand total. This is because the spools come in a finite length and if you order two 170-foot lines, totaling 340 feet, you could get a nice discount for buying more than 300 feet. If you bought a 300-foot spool, you would make one 170-foot line and then only have 130 feet leftover for something else. Ordering your cut lengths will ensure that you get your biggest discount without ending up with any short lengths of rope that you won't have much use for on your boat.

When you have calculated all your rope lengths, you should then write them all down as a list so you don't forget any of them while you are ordering them. The list should look something like this:

Rope	Diameter	Length	Color	Length Rounded Up
Main Halyard	9mm	125 feet		130 feet
Jib Halyard	9mm	129 feet		135 feet
Staysail Halyard	9mm	78 feet		85 feet
Trysail Halyard	9mm	125 feet		130 feet
Spinnaker Halyard	6mm	129 feet		135 feet
Main Sheet	9mm	172 feet		180 feet
Jib Sheet	9mm	95 feet (2 of them)		100 feet
Staysail Sheet	9mm	60 feet (2 of them)		65 feet
Spinnaker Sheet	6mm	101 feet (2 of them)		110 feet
Main Downhaul	6mm	50 feet		55 feet
Genoa Downhaul	6mm	66 feet		70 feet
Staysail Downhaul	6mm	34 feet		40 feet
Spinnaker Downhaul	6mm	66 feet		70 feet
Cunningham	7mm	21 feet		25 feet
Main 1st Reef Line	8mm	80 feet		100 feet
Main 2nd Reef Line	8mm	105 feet		120 feet
Main 3rd Reef Line	8mm	130 feet		140 feet
Dock Lines	1/2 inch	40 feet (8 of them)		50 feet
Main Topping Lift	9mm	133 feet		140 feet
Main Traveler	7mm	45 feet (2 of them)		50 feet

Rounding up will make sure that you don't end up short. If the line is too long, the excess can always be trimmed off, but you can't add to the line to make it a smidge longer. Rounding up will simply give you extra leeway in the event that your measurements were wrong or your calculations inaccurate. Having a little extra cushion is

SETTING UP THE RUNNING RIGGING

a welcome event on the boat and it will only cost a tiny bit more to ensure that you never work with the bitter end of a line because you ordered it too short.

REEVING SPLICE

All your halyards should have something called a reeving splice in the non-working end of the line. This loop is not structural but works wonders to help pull the line through by aid of a messenger line to install the new halyard. In the future, when you need to replace your halyard, you can easily tie a messenger line to this loop and pull the halyard out and the messenger line into place. Then doing the reverse, you can pull your new halyard back through and never have to worry about the messenger line knot slipping off the halyard's bitter end or, worse, the knot being too bulky and getting stuck somewhere. The reeving splice allows you to make a secure attachment with nothing more than a bowline. It's a small detail that really makes the difference when you need it.

INSTALLING A HALYARD WHEN OLD ONE IS PRESENT

Replacing an existing halyard is super easy to do if the current one has a reeving splice on the bitter end. If this is the case, simply tie your messenger line to the reeving splice using a sturdy bowline, tied with a rather long loop so that the knot doesn't get bound on anything as it makes its long journey. Once it's tied, simply pull the end with the shackle and watch as your halyard disappears into the mast and your knot pulls along right behind it. This is the moment when you really begin to question your knot-tying skills for if the bowline slips apart, you will need to skip the rest of this section and read the next section that discusses how to install a halyard when none is present. As long as you did a good job with your knot, the messenger line will chase the end of the halyard up the mast, over the shiv, and down the outside of the mast. As the halyard pulls farther along, you will switch your attention from pulling the halyard to resisting the pull of the messenger line as the messenger line will be the only thing acting to oppose the siphon effect that is pulling the halyard through on its last big journey.

Once the halyard is pulled all the way out and you have the messenger line running the entire route of the halyard, you simply need to untie the old halyard and tie the same good knot to the reeving splice of the new halyard as you are now going to do the same thing in reverse.

The reason you can't pull the halyard the other way is because the halyard shackle is just too bulky to fit through the gate on the side of the mast or through the small passage over the sheave in the mast truck.

If the person who bought the original halyard didn't have the forethought to have a reeving splice included at the bitter end of the halyard, you can still make do. The first option is to put a reeving splice in the end of the old line, but this is tedious as old line can be stiff to work. The other option is to stitch the messenger line through the rope by pushing a fid straight across the rope, piercing the cover and core and slipping the line right through the middle. Doing this a few times will create enough resistance that the messenger line can't slip out. Finishing the tail of the messenger line with a stout rolling hitch tied onto the old line will further the security of the connection. This method is as reliable as tying a bowline to a reeving splice, but it is much more time consuming, which is why the presence of a reeving splice is a welcome sight.

All of this can be done at the deck level, without ever going aloft, as long as you have a 1:1 halyard system. With a 2:1, you will need to go aloft to attach and secure the end of the halyard to the masthead casting's halyard attachment. If you have a 2:1 halyard, you will simply carry out the procedure like if you were a 1:1 halyard, but then fish the eye splice through the halyard block in preparation for your journey aloft. Pull out a mast's length of halyard and tie an alpine butterfly knot into the standing part of the halyard and attach your bosun's chair to the loop. The alpine butterfly knot will not damage the halyard and will also untie easily even after being heavily loaded. It provides a secure point for you to safely attach your chair and will never untie while in use. Having a very long tail is important because you are going to be attached to the line while you are trying to work. If you make the tail too short, you will realize the error of your ways when you want to come down. The tail needs to be long enough for you to come back down to the deck, otherwise the load holding you up will be switched from the halyard on the winch to the halyard on the anchor. The tail should be a little longer than the height of the mast, that way you can do your job and come back down.

An important safety issue, never attach your chair to the halyard block on a 2:1 system. If the halyard block fails, you will fall. If the anchor point fails, you will also fall. If the boat has a working 2:1 halyard and you are going up for any reason, tie an Alpine butterfly knot in the halyard and attach your chair to the loop. This will convert the halyard from a 2:1 to a 1:1 and will also keep you safe while you are suspended high above sea level.

If anything went wrong in the process and your halyard fell on the deck without bringing the messenger line with it, have no fear because you will learn how to install a new halyard in a mast that currently doesn't have an existing one to lead a messenger line.

INSTALLING A HALYARD WHEN OLD ONE IS NOT PRESENT

If you are installing a brand-new halyard in a position that you never had before, say a dedicated trysail halyard, or you had the misfortune of your halyard breaking and

SETTING UP THE RUNNING RIGGING

falling to the deck, you are in the same position as you will need to run the messenger line so that you can install the halyard like normal.

You might think that running a new or additional halyard is as simple as simply going to the top of the mast and pushing the bitter end over the sheave and down the spar, but this is not the case. A new rope will have bends and twists in it from its lifetime of shipping and storage and this means that your new halyard would get caught up and stuck on something inside your mast.

When routing your new halyard, you want the halyard to run a straight line from the sheave to the gate without tangling or fouling on anything inside the mast. The trouble is that you can't see in there and things are bound to wrap on each other. There is a trick to keeping things organized and it involves a heavy weight and magnetism.

Using a bicycle chain tied to the end of the messenger line, you can generate a large weight at the end of the line to help pull the messenger line through and over obstacles while also granting yourself the ability to control the path the chain takes with a magnet from outside the spar. Using a really strong magnet, the chain will be held firmly to the side of the mast where you can control it. Keeping the chain on the side of the mast closest to the sheave will ensure that the chain doesn't spin around in the mast and wrap around the other halyards or topping lift in there. If they do wrap, you will suffer from added friction when raising or lowering the sails, as well as notice additional chafe occurring on the section that lives inside the mast. This will shorten the lifespan of your halyard as you are inducing a chafe point, and one that is hard to identify. Wraps can occur despite our best efforts and without our knowledge, so it is always prudent to keep an eye out for any chafe on long sections of the halyard, particularly in strips along the length of the line. This will indicate to you that the halyard is being rubbed on something as it is being raised or lowered. The other line that is remaining static during these procedures will develop advanced and localized chafe on the spot that is contacting the halyard. If you see these long strips of chafe on a halyard, it is prudent to pull out all your lines in the mast and inspect them for chafe, as it is much easier to replace a halyard that is presently run than to install a new one from scratch.

If you do find that your halyards have wrapped inside the mast, you have two choices. The first is to ignore the problem and simply inspect the lines more frequently, replacing them as needed. The second is to untangle the lines by running new messenger lines for all the running rigging that runs through the mast with the chain and magnet in an attempt to correct the problem. If you are doing this, it would be best to run the messenger lines while the current halyards are still in place as you will need all of these halyards to get you up the mast to feed the messenger lines in. If you pull out all but one and then find a problem that precludes you from using that halyard, then you will find yourself in a situation where you have a mast with no safe way to get to the top to run new halyards and your only prudent method of action will be to have the mast removed by a crane so that the new halyards can be run at ground level, then have the mast re-stepped.

Back to the working part of this project: You hoist yourself to the top of the mast and then push the bicycle chain over the sheave and down into the mast. Once the chain has enough weight inside to pull the rest down, it is time to grab your magnet and place it on the side of your mast. When it catches the chain, you will hear the "clink" sound as the chain is pulled against the aluminum and that lets you know you have it!

As you lower yourself down, you will be paying out messenger line and pulling the chain down with the magnet. If your magnet looks like it might mar the surface of the spar, you can use a thin cloth to help protect the spar's finish as the magnet drags over the surface. The magnet won't be attracted to the metal of the mast because aluminum is non-magnetic, but your steel chain inside is magnetic and under your full telekinetic control.

On your descent, there are a few key things you want to focus on. First, the halyard should stay on the same side that it started. If the halyard sheave is on the port aft side of the mast, the halyard should run down the port aft side of the mast. If the halyard is on the starboard forward side of the mast, you guessed it, keep the chain on the forward and starboard side of the mast. The areas where this becomes important are when passing over the spreaders, as the compression post inside the mast can become a chafe point on the halyard. If your genoa halyard runs along the aft side of the spreader's compression post, at rest it won't matter but when the sail is raised, the halyard will now be routed behind the spreaders instead of in front of them.

A mast is normally raked aft slightly so this means that the aft halyards and topping lift are easy to route, the only thing you need to do is keep them organized. The line that uses the starboard sheave stays on the starboard side and the one on the port sheave stays on the port side. Gravity will keep the messenger line's chain pressed against the aft side of the mast so it will naturally slide behind the compression post. The headsail halyards will want to fall onto the back of the mast and slide behind the compression post and this will create a problem. When guiding the chain down the mast with the magnet, you not only need to focus on the port and starboard aspect of it all but also be mindful to make sure the chain passes over the front of the mast in the spreader areas. This will keep your halyards happy when they are in service and protect everything inside your mast.

If you have more than two sheaves on one side of the mast, simply run the most central messenger lines first, then tie them off and pull them tight so that you can run the next outboard messenger line.

SETTING UP THE RUNNING RIGGING

If the chain breaks away from the grasp of the magnet, you know that the chain will then swing around in the mast for a little bit. The pendulum force will cause it to sway and you know it would have created at least one wrap around another halyard in the mast! If this happens to you, your best option is to go back to the top and start again, but your other option is to be mindful of the event and keep an eye out for chafe. If chafe never occurs, then you got very lucky that day, but if chafe is occurring, you will know why and also how to fix it.

The other reason for using a magnet and chain is because it makes pulling the messenger line out of the gate so much easier. Instead of fishing for it by blindly feeling around inside the mast or hoping to snag it with a hooked wire, you can precisely drag the chain right over to the gate, removing the barrier from their attraction and letting the magnet and chain finally meet in intimate contact. You now have a firm grasp on the chain and can pull it out to expose the messenger line. Running a new messenger line can be easy as pie if you use a bicycle chain and magnet. Once you have the messenger line present, you can then tie the end to the reeving splice and pull the actual halyard down through the spar and out the gate.

If you have external halyards, none of this is to be a concern to you as you simply need to feed the new halyard over the sheave where that halyard will be working and then across the top of the masthead casting to the other side where there will be another sheave that will guide it down again. The mainsail halyard will enter on the back and exit on the front, while the jib halyard will enter on the front and exit on the back.

CHAPTER NINETEEN
Tune Your Sailboat

Tuning your standing rigging is much like tuning a piano: there are a ton of small components that all need to work in absolute harmony to create a beautiful melody and if anything is even slightly off, everything is awful.

The first goal of standing rigging is to keep the mast in column. This means having the mast be perfectly straight so that every section stands directly over the previous one. Picture a small child stacking blocks, if the child places each subsequent block perfectly over the previous block, the tower could be as tall as the ceiling height allows. If the child is stacking the blocks in a haphazard way, then they will be lucky to make a tower of even just a few blocks.

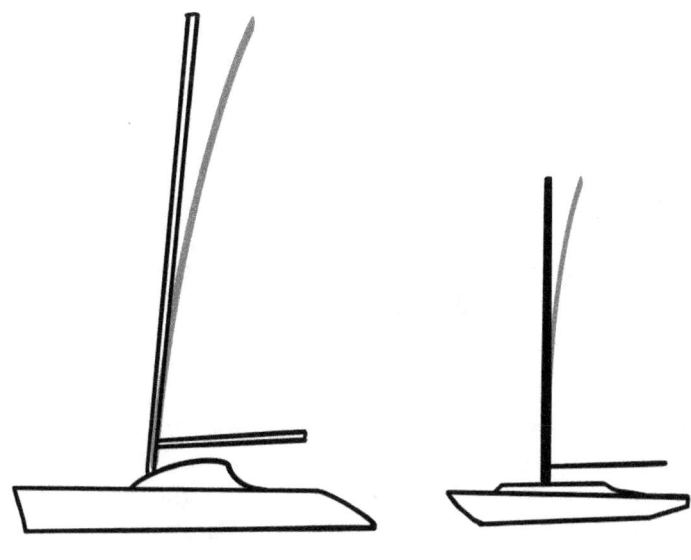

A slight aft curve and rake is ideal, the only exception being a boat with in-mast furling as these need to be straight up and down with no curve or bend to them at all. Any curve will cause the furled sail to bind against the internal walls of the mast and make it get stuck.

If the mast step is situated too far aft, the mast will have a very pronounced aft curve and the way to remedy this is to move the mast step forward. If the mast has an S curve when viewed from the side, then the mast step is too far forward and the remedy will be to bring the mast step aft. Both of these are no easy feat as moving the mast step is not normally an easy job. Some boats boast adjustable mast steps, but these are few and far between. Normally the mast step is bolted to a support that is integral to the hull or deck, meaning that these adjustments are going to be both very involved and expensive.

If the mast appears to be bending forward, this is called an inversion and is a terrible thing to see on your boat. Inversions are caused by a variety of issues relating to the rig tension. They can stem from insufficient backstay tension on boats with in-line spreaders, insufficient cap shroud tension on boats with aft swept spreaders, as well as too much wind pressure on the sails with a loose rig. An inversion is a negative bend in the area of the spreaders. The mast section is not designed to bend in this way and when it happens, the mast can buckle and break. The middle moves aft while the masthead gets pushed forward, and this is a very dangerous situation to be in, all caused by rigging that is not tight enough for the spar's needs.

When you look at the mast from the front, you want it to appear perfectly straight, from the side you want it to have a gentle aft curve. Anything deviating from this will require you to look at the original plans for your boat and make sure that your current rig appearance matches the drawings. High-performance racing yachts will have a significant aft bend in the mast, and schooners will have very pronounced aft rake in the masts. This is not wrong, it just is not the norm. Having an uncommon setup doesn't mean that your setup is wrong, but simply that you need to make sure that your setup is unique and not in need of correction.

There are two ways to tune your mast and get it perfectly in column: One is to work from the bottom up, and the other is to work from the top down. When working from the bottom up, you need to center the lowest section of the mast, and then center the next section, and then center the next section, until you reach the masthead where you will center it over the boat and also over all the sections that came before it. The problem with this method is that if you didn't center the lowest section perfectly, you will end up tuning your mast perfectly in column but off to the side. Any slight imperfection with the lowest section will then lead to the compounding of these errors moving upward. At the very end, you can then hang a plumb weight from the main halyard to verify that the masthead is centered over the boat, only to find that it is a few inches or feet off and the entire process needs to be repeated again.

Alternatively, you can tune from the top down where you start by centering the masthead directly over the boat. Naturally, this is done in the water with the boat floating, so the boat needs to be trimmed level. If your boat always has a slight list to port, you will need to throw some weight on the starboard rail to trim the boat out. If your boat is very tender, you can just adjust where you stand to trim the boat, but do so in a way that you can see your level gauge to know when you need to stop moving.

With the boat trim, you will simply hang a heavy weight from the main halyard and position it just above the gooseneck. The goal of this exercise is to have the weight hang perfectly behind the mast in perfect line with the mast track on the back of the mast when viewed from directly behind. If the boat is level but the mast is off to the port side, the weight will hang to the port side of the mast. Once you have the masthead centered directly over the boat, you know that you can tune the rest of the mast visually to bring it into column and the mast will be perfectly centered over the boat. That means when you're done tuning the mast, you are actually done tuning the mast.

The weight behind the mast track works to get the masthead positioned athwartship, but now we need to position the masthead in relation to the mast step, getting the perfect amount of rake. If you have no drawings or plans available and you are doing the first tune of the boat, the general starting point is to have the head of the mast positioned just aft of the heel of the mast. In practical terms, you will lower the weighted halyard so that the weight now hangs just above the deck and the halyard rests next to the boom. The halyard should cross the boom "one mast's distance" behind the back of the mast, measured at the gooseneck and boom. If your mast is 4 inches fore-aft, then the halyard will hang 4 inches behind the mast. If you have a bigger mast that measures 16 inches fore-aft, then the halyard will hang 16 inches behind the mast. This might sound like a drastic amount of rake, but as the mast gets thicker it also gets longer and the amount of rake is about the same. This is just a starting point, and if you find that the boat has too much weather helm, you will reduce this dimension and move the masthead slightly forward. If you have too much lee helm, you will do the opposite and move the masthead back a little more. Small incremental adjustments where only one variable is changed with a test sail to evaluate are better than doing drastic changes and not knowing what actually helped.

What can you use for a weight? Anything that is heavy enough to hold the halyard tight and straight that you have on the boat. My preferred method is to use a full jerry can, as nothing says safety more than 40 pounds of flammable liquid swinging around the boat. This should serve as a reminder that you will only do this on a calm and windless day, that way the boat is at rest and stationary. If you have teenagers zipping by on jet-skis or ferry boat wakes rolling through, then you will need to find a better place to carry out this exercise. Calm surface waters are not the only thing you need to consider, as the air also needs to be still. There is no point suspending a weight if the windage will alter the result of the test.

While tuning from bottom to top or top to bottom, you will need to sight the mast to see if you're putting the mast in column. To sight the mast, you will look up the mast with your chin against the front of the mast. This will make the front of the mast look like a long road ahead. If the road has twists and turns, then you have some work ahead of you. If the road is straight, then you are looking good! Sighting the front of the mast will let you see how the mast is tuned in an athwartship dimension; to see how the mast is behaving in a fore-aft dimension, you will do the same but from the side of the mast. If the view from the front is obscured by equipment, you can also look up the back of the mast, following the nice black line of the mainsail track.

When sighting the side of the mast, you want to see a slight curve aft, unless you have in-mast furling in which case you want the mast to be perfectly straight in all dimensions. When you sight the sides, if you see on the port side that it looks like it's coming toward you and on the starboard side that it looks like it's bending away from you, then you should look again from the front knowing this, and see if it looks like the mast is curving to port. Sighting the front is a great way to get a general sense of the spar's direction, but sighting the sides helps confirm and corroborate these findings. Think of it as checking your work with a new perspective.

While tightening your cap shrouds to position the masthead, it is time to take an important measurement if you have steel rigging with turnbuckles. This measurement is the amount of stretch that occurs with each turn of the turnbuckle to determine how much tension you are actually applying to your rigging. This is only really important for rigs with aft swept spreaders as they need to have a minimum of 25% of the breaking strength on their cap shrouds, otherwise they will have insufficient forward push on their spreaders to give the mast its gentle bend as the wind builds. Without this forward push, the mast can invert and possibly collapse. For masts with aft swept spreaders, this minimum must be respected and you therefore have to do this tedious but simple calculation. For masts with in-line spreaders, tension is less critical as the shape of the mast is derived directly by the rigging so as long as the mast has the appropriate shape, you can run a slacker rig and still get away with it.

1×19 Stainless Steel Rigging, both 304 and 316, will stretch 1mm over a 2m length for every 5% of its breaking strength that is applied. This means that if your rigging needs to be tensioned to 25% of its breaking strength, then you need to see that the wire will stretch 5mm as you tension it. The easiest way to measure this is to use a simple method that is known as the Folding Rule Method.

The Folding Rule Method uses a 2m long ruler attached at the top side to the stay with the bottom side set 1mm above the end of the wire of the stay, as it enters the terminal. As you tighten the rigging, this small gap between the terminal and the end of the 2m ruler will widen and this distance is directly correlated to the stretch and therefore the tension on the wire. The cap shrouds are linked at the top of the mast and therefore any tension placed on the port cap shroud will be equal to the tension on the starboard cap shroud. As you turn the port cap shroud turnbuckle, you will be applying tension to the port and starboard cap shrouds.

To start the procedure, you will attach the top of a 2m ruler with the bottom just 1mm above the terminator fitting at the end of the wire of the starboard cap shroud; this can be measured with an accurate digital Vernier caliper. Turn the starboard cap shroud until the wire starts to become tighter and the gap has now increased to 1.5mm. At this point, you know you are in business and the wire is starting to stretch. Now go turn the port cap shroud turnbuckle until the gap widens to 3mm on the starboard cap shroud. Now you know that both are in tension and the rigging is taking on a load. Now, count the number of turns you needed to do on the starboard cap shroud turnbuckle to bring the gap from 3mm to 4mm. This +1mm will correspond to the amount of turns it takes to add 5% of the breaking strength to your rigging. Write this number down because it is very important to keep for the duration of your ownership of that vessel. Anytime you need to tune your rigging, you just need to count your turns, knowing the number of turns per 5% of breaking strength. This value only works on wire that is that exact diameter so if your lowers, intermediate stays, or backstays are of a different diameter, you will need to carry out the same test on them as well. If your rigging is all the same diameter, then you are done with this testing as you now know the number of turns for 5% of your wire's breaking strength.

If you need to tension your rigging to 15%, then you will simply do that number of turns three times; if you need 20%, you will do it four times, and if you need 25%, you will do it five times. This magical value will ensure that you always provide your rigging with enough tension while not exceeding the safe limits for your rigging.

With the masthead centered and your number of turns determined, you can then begin tuning your rigging. First, you will center the masthead so that it is directly over the heel of the mast that rests on the mast step. Second, you will turn the turnbuckles to the point where they will supply the needed tension. Knowing that every set of turns will increase the tension on both cap shrouds, you don't want to do your number of turns on both sides because that would give you double the tension desired. If you know that four turns makes 5%, and you want 20% of the breaking strength then you know you need to do 16 turns total. This means eight turns for the port cap shroud and eight turns for the starboard cap shroud. If you did 16 turns on each, you

THE RIGGING HANDBOOK

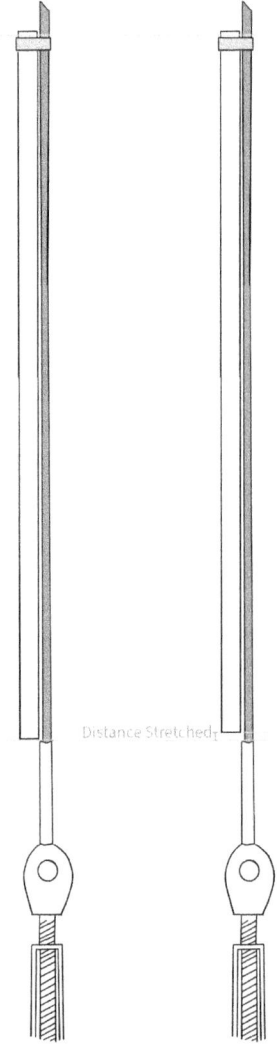

would have the rigging tensioned to 40% of the breaking strength, which is far in excess of the safe limits for the standing rigging.

Once the masthead is centered and tensioned, your next station to work on will be the next set of shrouds on your way down. If you have intermediate stays, you will then need to tension them to bring that section into column. S bends in the mast are due to the intermediates and lowers being out of whack with each other and will require their immediate adjustment. Remembering that a stay will only pull, will help you straighten out these issues.

Tightening the stays that are on the concave side of the bend will pull the mast back into column. The only time you want to loosen a stay during adjustment is if the stays feel like they are too tight and that you might have exceeded the maximum tension.

If you have a multiple spreader rig, you can develop an S bend or C bend, which can be easily tuned out. If you have a single spreader rig, you can develop the C bend, which is easily tuned out by tightening the lower on the concave side. This will pull the mast over to the correct side and bring everything back into column.

While the Folding Rule Method is critical for situations where you need to have an exact tension on your rigging, there is another method for in-line spreader rigs that do not have this strict requirement. Mast manufacturers will indicate that you should tension your rigging with in-line spreaders to a maximum of 15% of the wire's breaking strength. The truth is, they just need to be "tight enough." 15% feels pretty tight when you give the stay a jiggle, and if you feel enough stays that are set up properly on other boats, you can actually develop a "feel" for how tight is tight enough. Once you have the feel, or you used the Folding Rule Method for the cap shrouds, the rest can all be done by feel. On in-line spreader rigs, the order of tightness from tightest to loosest is:

1. Cap shroud

2. Intermediates (if present)

3. Forward lower

4. Aft lower

Once you have the tension and position set on the cap shroud, you will simply set the tension on your intermediates to straighten the mast out at that section while making sure that the tension does not exceed the tightness during the jiggle test of the cap shroud. Once that is done, move on to the forward lower where the tension should not exceed that of the cap shroud or intermediate, and should straighten out the lower portion of the mast while also giving the mast a gentle prebend. Lastly, the aft lower will help finish off the athwartship position of the mast and be looser than the forward lower. Once the tuning is completed, simply give the stays a jiggle in that order and make sure that the tension on them feels like it is in a decreasing order. If any stay feels tighter than the previous stay, then you can look at what the mast is doing and why that is the case. The problem should be rectified by either loosening the stay that is too tight or tightening the stays that should be tighter, as they might have been too loose.

The rig tune at this point is referred to as the "Dockside Rig Tune," which involves the boat sitting in the slip and tuning the rigging. If you try tuning the rigging while on the hard with the boat on jack stands, you will find that the rigging will be completely out of tune when you start floating, and this is because the hull deforms when it is on the hard and supported by only a few key points. In the water, the boat is supported differently and therefore changes shape slightly. By now you know how important every fraction of a millimeter is, and why this slight change in the hull will spell disaster for a perfect tune.

The dockside tune done in the water will need to sit for a few days, as the load will pull on the chainplates, which in turn will pull on the hull. This tension will cause a permanent deformation in the boat's hull, known as creep. Creep is different from stretch because creep is the permanent elongation of a material whereas stretch is non-permanent. When the load is removed, something that stretched will go back while something that creeped will remain longer than it was before. The elongation of stainless-steel wire rigging as you crank on the turnbuckles is stretch because if you loosen the turnbuckles, it will un-stretch.

Older boats will deform less from the loads of new rigging because they have already done their creep and have settled into their shape. Brand-new boats are going to creep for a while and this will cause the rigging to become very out of tune. This is not a problem, but more a phenomenon that needs to be dealt with and accommodated. All this means is that after the dockside tune, the boat needs to rest for a few days and then check to see if the rigging has settled. If the rigging is out of tune after a few days, then the dockside tune needs to be performed again, and again allowed to rest. Once the rigging has settled into place and the tune holds, the next and real tune can take place. This is the tune that is performed under sail.

The entire purpose of tuning the rigging is so that the boat can hold the sails into the breeze. How accurate is the tune if it is not checked under the very conditions and purpose it was dedicated to do? After a mechanic performs work on your car, they will then take it for a test drive to make sure everything worked out well.

Tuning under sail is very simple, and actually easier to do than the dockside tune. When performing a dockside tune, you are adjusting the rigging in such a way that it "should" work when under sail. There is a lot of guesswork and estimation to assume that this is the correct amount of tension for a particular stay. What if the step down in tension between each stay is too drastic and now the mast dips to leeward too much in the middle. Dockside tuning is something that takes a lot of experience to get right, but that doesn't mean that you can't do a good job if you don't have experience; you simply need to check your work with the sail tune.

To properly tune the rig under sail, you want to be on a close reach with the boat under full sail and heeling around 20°. If you are heeling closer to 30°, then the wind is too strong and you need to try again on a better day. With the boat leaning 20°, you will be loading all the windward shrouds well enough to see how they are performing together. You will sight the mast from the front as well as from the sides, looking to see if the mast is in column and with a gentle bend aft. You might not have this, but that is what you want to see. Instead, you might see that the mast is a bit wonky and that will identify where you need to adjust. The best part about tuning the rigging while under sail is you can see the direct effect of your work while you are doing it. While under sail, you can see the mast straightening out as you turn the turnbuckle and you know you are done when the mast is straight. This instant feedback is my favorite part about the whole process as you can see exactly how much is needed and you know it is perfect when you are finished. When you get back to the dock, you can calibrate your hands by giving the rigging a jiggle again to see how tight it is now that it's perfect. With time and experience, you will get better at the jiggle test until you can use it reliably.

If you have a double spreader rig, you will have the joy of dealing with S bends. These can be a little tricky to sort out because they can be caused by three different stays. If the cap shroud is too loose, the top of the mast will fall to leeward, the top spreader will appear to be pulled to windward, and the bottom spreader will react and slip to leeward. The remedy is to either tighten the cap shroud, loosen the intermediate, and/or tighten the lower. How do you know which you need to do?

The clue resides on the leeward side. If the windward cap shroud is too loose, then the leeward cap shroud will be flapping in the breeze. During the dockside tune, you would have centered the masthead and, as a result, both cap shrouds were tight and centered. While heeling over, if the windward cap shroud is tight but the leeward cap shroud is loose, that means that the head of the mast has shifted laterally and is no longer directly centered over the boat. The solution here is to tighten the windward cap shroud until the leeward cap shroud stops flapping in the breeze. You know the shrouds are "tight enough" but not too tight when the leeward shrouds go slack at 25 knots while under full sail on a close reach. If they go slack before, then you know you have loose rigging. If the leeward cap shroud is not loose, then that means that the cap shrouds are not to blame and the problem is in the intermediate or lower stays.

If the S bend is not caused by the cap shrouds, then you should check the intermediate stays next and in the same fashion. Give the leeward shrouds a jiggle and see if the intermediate stay is still tight, if it is, then this will tell you that the windward intermediate is too tight and not letting the mast move appropriately. If the intermediate was the culprit, then the S bend will resolve as you ease the intermediate and allow that section of the mast to move under the head of the mast.

If the mast now has a C bend in the bottom, and the leeward lowers feel incredibly loose, then you simply need to tighten the windward lowers to bring the mast over at the bottom spreader.

The order of operations is simple for S bends. If the leeward cap shroud is loose, tighten the windward cap shroud. Then check the leeward intermediate, and if it is tight, the windward intermediate needs to be loosened. Then check the lowers and if they are loose, the windward lowers need to be tightened. Working from top down, you turn the procedure into a systematic process where the problem will be removed in one go.

The order of operations for a reverse S bend (2) is as follows: The cap shroud can be ignored as it is not a culprit if the mast was centered properly, so start with the leeward intermediate and see if it is too loose. If so, the windward intermediate needs to be tightened. After that, check the leeward lowers and see if they are too tight; if so, the windward lowers need to be loosened.

Once you have your stays tightened to perfection, you will have the joys of tacking and then starting all over again because the other side needs the same attention. If you do not tune on both tacks, you will find that your boat will perform better on one tack than the other and this is because the mast is in tune on the better tack, and thus the sails are shaped correctly, while out of tune on the tack that doesn't perform as well, because the sails are not shaped correctly.

Single spreader rigs are much easier to tune as they are less dependent on the rigging to support them. This means that there is less to adjust and, as a result, less to go wrong and less to need to correct. S bends are not an issue because you don't have three points of attachment aloft to generate the S bend. If you look at it very closely, you can convince yourself that you are looking at a slight S bend, but what is really happening is you have a C bend with a little bit of flair at the bottom. The mast is supported at the top and in the middle, so one of these two places will be loose and need to be tightened.

If the cap shroud is too loose, the leeward cap shroud will also be flapping in the breeze, to which you will simply need to tighten the windward turnbuckle until the mast is in column and the leeward cap shroud shores up a bit. If the lowers are too loose, then the middle of the mast will bow to leeward as the lowers are not providing enough tension to oppose the compressive force generated by the spreader from the cap shroud being pushed out to the side. Tightening the lowers until the mast is in column and the leeward lowers shore up a bit will be all that is required to fix the situation. Once the mast is perfectly in column, you will simply tack and repeat the process on the other side.

Turning turnbuckles is really easy to do with the right tools, all you need are two open end wrenches. The smaller one will go on the lower terminator just above the threaded portion, as there will be two small flat surfaces that will fit a specific sized wrench. This wrench doesn't move and simply keeps the top wire from spinning as you turn the turnbuckle. The second wrench will be bigger and will slide over the turnbuckle body. The leverage between the two wrenches will give you plenty of mechanical advantage to tighten a turnbuckle, no matter the load that is placed on the rigging while you are sailing on a close reach.

A little caveat with cutter rigged sailboats that have a single spreader, as these can develop S bends in them because the inner forestay is opposed by two backstays. These can be running backstays that go aft and give the mast wonderful support to oppose the forward pull of the staysail on the mast or they can be static attaching to the boat just aft of the aft lower chainplate. These have less ability to oppose a large staysail genoa as their staying angle makes them act more as a shroud than a backstay. If these stays are too tight, they can cause the mast to have an S bend even though the mast only has one set of spreaders and should only get C bends.

These rigs need to be treated the same as a double spreader rig when it comes to determining which stay needs to be tightened or loosened. If the leeward stay is too loose, the windward stay needs tightening. If the leeward stay is just loose, the windward stay is fine and the other stays need to be adjusted instead.

After a sail tune, your boat should hold its tune for the entire sailing season. If you are hyper focused on performance, then you should have your rigging tuned every season, but if you are a relaxed cruiser, simply sight your mast while you are underway to make sure that everything is copacetic. If everything looks good enough, then you can keep on sailing and focus your time on all the other aspects of your boat that do need your attention. If you see that the mast is not in column and you have the time because you are on a long passage with nothing else to do, then you can tune the rigging and give you something to talk about at the next watch change. If you don't have the time to address it now, then keep it in mind and take care of it as soon as you can. Knowing that the mast is not in column will also influence how you treat your rigging while cruising. If you are sailing in a gale and know that the starboard side is fine but the port shrouds are a little off, then attempting to ride out the storm on starboard tack will be preferable as this is the boat's good side. Knowing that the mast is not in column will also motivate you to reef early and attempt to keep heeling to a minimum as you know that the more you are heeling the more force you are putting on the mast and rigging. Keeping the loads down will also reduce the amount of punishment that you are subjecting the spar to until the storm has passed and you can then remedy the situation.

All of this so far has focused on turnbuckles, but what about synthetic rigging that doesn't employ a turnbuckle? Can you do this same stuff with synthetic standing rigging and deadeyes? Yes you can! The only thing you can't do is "count the number of turns of the turnbuckle" to determine what is 5% of the breaking strength. Tuning synthetic standing rigging is done completely by feel and then corroborated by a tune under sail. Deadeyes take a long time to set up and tie off, so this will take a lot longer than it would with turnbuckles. This means that you should avoid doing this in narrow waterways where you will run out of water before you finish your work. Try to find somewhere that you can be on the same tack for a few hours; that way you will have less traffic as well as less distraction while you are focusing on your rigging.

This next part will depend solely on your confidence in your ability to tie knots. Dyneema is a very slippery material and knots that are not tied well will pull out. As soon as you remove the shroud frapping knot from the lashings, you are solely dependent on the knot you tied to connect the tails to the tensioning line. If this knot slips out, the lashings will pull free and your stay will go with it. If you have any doubt in your ability to tie a good knot in Dyneema, especially with the cap shroud, then don't tension the stay while it is under load. Simply tighten it while it is on the leeward side and then tack to check your work.

While Dyneema will pull if not tied correctly, tying a very long tail that will give you notice to the predicament as well as stop knots in the ends of these long tails will give you advanced notice that a problem is starting to develop. I personally try to do the final tensioning while under sail, that way I can tighten the stay until the mast is perfect, and then I can tie it off knowing that the job is done. My preferred knot for this process is a sheet bend with a slip. The slip allows easy untying after the job is done and the long tails, as in several feet long, give ample warning that the knot is slipping. Loading the knot slowly will help it seat fully and prevent any slippage, but if some should occur, you can always try to push on the knot to help guide it into a locked position where the slipping stops. If you can't get the slipping to stop, you can tack to take the load off of this stay and give you time to sort out the issue and try again.

If you are cruising full time, crossing oceans when the weather is right and island hopping through paradise, you can expect to get a few years out of your rig tune before small imperfections will start to show up. If you are a racer who retired to cruising but still has that nasty addiction to speed, then you will need much more frequent rig tunes.

Synthetic rigging can hold a tune for years if you stick to your temperature range and avoid sailing in temperatures that are outside of the +/-20°F window that you tuned your rigging for. From personal experience, we sailed about 16,000 nautical miles with two transatlantic voyages and lots of coastal explorations on four continents over a three-year period without a single rig tune. As temperatures fluctuated, the stays would expand and contract, but all at the same rate. The cap shrouds would get longer than the lowers, but at a proportional rate so that the entire mast simply leaned a little bit more to leeward in weather that was at the cooler end of the spectrum for our rig tune. Knowing the rigging was slack as the temperature dropped, we simply kept the sails lower to reduce the load that was placed on the rigging. When temperatures warmed back up, the rigging would contract again, but all at the same rate so that the mast simply remained in the place it was intended. This worked because we were relaxed about our rigging and simply made sure to not push it when it wasn't at 100%. Over time, you will start to see that the mast isn't perfectly in column anymore and that is your cue that the rigging needs a tune. Naturally, this relaxed approach worked for us because we have in-line spreaders and can get away with a loose rig. If we had aft swept spreaders, we would not have been able to let the rigging go this slack as the mast would probably not have survived the ordeal.

Inversions are a scary thing on a sailboat. The mast is designed to bend aft slightly or be perfectly straight. Any bend forward is beyond the design limitations of the spar and catastrophe is right around the corner. Things that can make an inversion a little more possible are loose rigging, particularly a loose backstay as this will let the top of the mast pull forward. Another big factor that can further this destructive form is to reef the mainsail lower than the headstay, as this will cause an aft pull on the mast with minimal opposition provided by the rigging. The mainsail will pull evenly along the luff, and if all that pulling is concentrated in the middle of the mast, then that part is going to be pulled backwards. This compounded by a loose backstay and a fractional headstay is a recipe for disaster!

Having a cutter with an inner forestay or a sloop with a baby stay are great ways to prevent inversions because they will pull the middle portion of the mast forward and oppose the force from the backstay, inducing a luscious prebend in the mast that can counter the force of a reefed mainsail in its attempts to pull the mast back and spell destruction.

Afterword

This book is only the first step in your rigging educational journey. I encourage you to continue your studies through literature, but also to apply what you have learned here by giving your own rigging a proper analysis.

Hopefully by now you feel equipped with a basic understanding of the global effects of rigging on a boat. You are now armed with the tools to apply this knowledge to your own specific vessel. Just as a professional rigger takes years to master the craft, so should you continue to learn and gain experience. It will take trial and error to gain confidence, but if you are determined and patient, you can absolutely inspect your own rigging, fix any issues, and even rerig the entire boat in the same or a different material. If you want to be a professional rigger, this book has served as a foundation to make you competent and proficient at selecting the right materials and sizing them appropriately, starting you out on the right track as you go on to gain practical experience over the years.

At the end of the day, sailboats are just a massive physics experiment—and you now know the formulas to unlock the answers!

Glossary

Aft. Toward the stern of the vessel.

Aft swept spreader. Spreaders that extend laterally as well as abaft. The force on the cap shrouds and intermediates helps push the mast forward in the middle. These spreaders are a sign that this rig is very tension dependent.

Aluminum. Most common material used for spars on the modern-day sailboat. Very strong and lightweight compared to wooden spars.

Anchor rode. Material used to connect the vessel to the anchor. Typically made of rope, chain, or a combination of rope and chain.

Angle of attack. Angle that the sail meets the wind. This is a critical factor in lift generation.

Backstay. Stay that runs from the back of the boat to the top of the mast and keeps the mast from falling forward.

Balanced helm. Sailboat's tendency to sail straight when the rudder is maintained straight.

Ballast. Weight set low in the boat to lower the center of gravity of a boat. This weight counteracts the weight aloft as well as the pressure of the wind on the sails and helps prevent the boat from capsizing.

Batten. A rigid structure inserted into the sail cloth to assist the sail in holding its shape.

Barque. A tall ship with three or more masts where the last mast is fore-aft rigged and all masts forward are square rigged.

Barquentine. A tall ship with three or more masts where the first mast is square rigged and all masts aft are fore-aft rigged.

GLOSSARY

Bobstay. The stay that extends from the end of the bowsprit down to the stem of the vessel.

Boom. Spar that runs below the foot of the sail and grants greater trim potential by positioning the clew in a more controlled manner.

Brig. A tall ship with two masts where both masts are square rigged.

Brigantine. A tall ship with two masts where the forward mast is square rigged and the aft mast is fore-aft rigged.

Bronze. Alloy used on sailboats due to its strength and corrosion resistance. Common alloys are aluminum bronze, silicon bronze, and manganese bronze, though only silicon bronze should be used in marine applications.

Carbon fiber. Composite material of incredible strength with minimal weight.

Center of Effort (CE). Geometric center of a sail.

Center of Lateral Resistance (CLR). Geometric center of the underwater profile of a boat.

Chafe. Damage to the surface of a material caused by rubbing.

Chain plates. Metal or composite components used to distribute the forces of the rigging to the hull of the boat. *Sizing:* based on the tensile strength of the selected material to support the rig loads that will be placed on that specific stay. *Material:* aluminum, bronze, steel, titanium, and carbon fiber are common materials used to manufacture chainplates.

Chord. Fore-aft length of a sail.

Clevis pin. Small metal dowel with a large head on one end and a hole for a retaining pin on the other end. Used to connect various rigging components that will be in tension. Makes for a fast and easy method of safely securing various components with minimal tools.

Clew. The after lower corner of a sail.

Close hauled. A sailing vessel is close hauled when she is sailing as close to the wind as she will go.

GLOSSARY

Coefficient of thermal expansion. Rate that a material changes size as a result of a change in temperature.

Continuous rig. Rig setup where the shrouds run continuously from chainplate to their specific point on the mast.

Cotter pin. Retaining pin used to secure a clevis pin and prevent it from working its way out of a fitting. In sailboats, the legs should only be bent 10°. If the legs will become a hazard and need to be bent more, that is your clue that a cotter pin is the incorrect retainer in this application.

Creep. Permanent elongation of a material.

Crevice corrosion. A type of corrosion that occurs in oxygen-poor environments and is very hard to identify.

Cringle. A metal hole in a sail.

Cunningham. Running rigging used to pull down on the luff. Useful for controlling the position of the draft of a sail.

Cutter. A fore-and-aft rigged vessel with a single mast usually stepped about two-fifths of the distance from the bow with two headsails.

Deadeye. Component used to receive the lashings that tension the stay and connect to the top of the chainplate.

Diagonals. Shrouds that run athwartship from the tip of one spreader to the base of another.

Diamond stay. A type of stay that originates and terminates on the same spar and is held out by a spreader.

Dissimilar metals. Two different metals that are used in close proximity and have different electrochemical potentials. The intimate association of dissimilar metals in the presence of an electrolyte will lead to galvanic corrosion.

Dock lines. Ropes used to secure a vessel to a pier or quay. *Material:* usually rope but can also be rope/chain combinations. *Size:* dependent on the length of the vessel. *Length:* dependent on the length of the vessel.

Downhaul. Running rigging tied to the head of a sail that is used to aid in bringing the sail down.

GLOSSARY

Draft. Depth of the curve of a sail, which is critical for power generation.

Dyneema. Polyethylene rope that can be used in standing and running rigging applications.
    DM20. High-strength fiber with no creep.
    Heat Set. Dyneema that is pre-stretched in a heated environment to make the rope much stronger and more resistant to creep.
    SK75. High-strength fiber with high creep.
    SK78. High-strength fiber with low creep.

Fisherman sail. Sail set to fill the space at the top between the foremast and mainmast on a schooner.

Foot. Bottom edge of a sail.

Foremast. Smaller mast that is stepped forward of the mainmast.

Foresail. Sail set aft of the foremast.

Freeboard. That part of a vessel's side that extends above the water.

Full-Rigged Ship or Fully Rigged Ship. A tall ship with three or more masts where all the masts are square rigged.

Galvanic corrosion. Electrochemical corrosion caused by dissimilar metals in contact while in the presence of an electrolytic solution.

Galvanized steel. Zinc-coated steel that is used to protect the underlying steel from the marine environment and delaying corrosion.

Gollywobbler. Sail set to fully fill the gap between the foremast and mainmast on a schooner, used in light wind situations.

Grommet. A rope ring.

Halyard. The rope used to hoist a sail.
    1:1. Halyard that is connected to the top of the sail directly and raises the sail by the equivalent length of halyard that has been moved.
    2:1. Halyard that uses a pulley system at the top of the sail and raises the sail by half the length of halyard that has been moved. Common in applications where the sail is very heavy and the force multiplier is needed to lift the sail.
    Halyard wrap. A dangerous situation where the halyard wraps around the headstay; commonly occurs with roller furling when the halyard and headstay run

GLOSSARY

parallel to each other. If the wrap is ignored, the stay can break and cause the mast to fall.

Head. Top corner of a sail.

Headsail. Sail set on the headstay.

Headstay. Stay that runs from the bow to the top, or close to the top of the mast. This stay keeps the mast from falling backwards and is used to attach a jib or genoa.

Hollow Leech. The concave curve of the leech of a sail due to a lack of battens.

In-line spreaders. Spreaders that extend straight out to the side. These are used on rigs that require lower tension on the stays.

Inner forestay. Second stay that runs inside the triangle made by the headstay, mast, and deck. Usually found on cutters, but can also be found on other rigs as well.

Jib. Triangular sail set forward of the mast.
    Blade Jib. Type of jib with minimal draft; best for upwind use.
    Flying Jib. Jib that is set without being connected to a stay.
    Genoa. Type of jib where foot extends beyond the mast; best for upwind use.
    Yankee. Type of jib where leech and foot are nearly the same length; best used for reaching

Jib sheet. Running rigging used to trim the jib.

Jibe or Gybe. To come about from one tack to another without heading into the wind, a dangerous operation if carelessly done or if the wind is strong, as the wind striking the sail from aft throws it over with a jerk.

Ketch. A fore-and-aft rigged vessel with two masts, the larger of which is forward. It is distinguished from a yawl by the fact that its mizzen mast is stepped forward of the rudder post.

Knot. Method of connecting a rope to something via turns and friction. Usually results in a loss of strength of the rope.

Lashing. Thin line used to tension a stay or shroud. Held in place via the shroud frapping knot.

Lazy. Component that is not working at the moment, usually on the leeward side of the vessel.

GLOSSARY

Lee. The side of a vessel away from the wind, opposite to the weather side.

Leech. The after edge of a sail.

Leeward, or loo'ard. The direction toward which the wind is blowing.

Lee helm. Sailboat's tendency to turn downwind if the rudder is maintained straight.

Line. Rope that is in use on a boat. There are several exceptions but, in general, a rope is for sale on a spool at a chandler and it becomes a line once it is installed on a sailboat.

Lines led aft. Running rigging is led to the cockpit, allowing control of the sails to be managed from the cockpit.

Linked rig. Rig setup where the shrouds run in sections from chainplate to spreader tip and terminate there. A different shroud then begins at the spreader tip and runs farther up the mast to the next terminating point. The shrouds are made up of various individual pieces that are linked together.

Long splice. A non-locking eye splice used with single braid ropes. The resulting splice is stronger than a Mobius Brummel splice, but has the ability to work itself free when not under load.

Luff. The forward edge of a sail.

Mainsail. Sail set aft of the main mast.

Mainsheet. Running rigging used to manipulate the position of the boom and to set the twist in the mainsail.

Mast. Spar that supports the sails in a vertical dimension.

Mast step. Specialized structure that supports the bottom of the mast and prevents it from sliding or moving.

Mechanical advantage. The increase in force that is generated by the use of simple machines.

Minimum staying angle. 12.5°. If the stay approaches the spar at less than this angle, it will not be effective and will simply result in unnecessary strain on the other components of the boat and rigging.

GLOSSARY

Mizzen. Smaller mast set aft of the mainmast.

Mizzen sail. Sail set aft of the mizzen mast.

Mobius Brummel splice. A locking eye splice used with single braid ropes. Suitable for manufacturing eye splices in standing rigging.

Nylon. Plastic fiber that has high elasticity. Ropes made from these fibers will be very elastic. Strength is greatly reduced when nylon ropes get wet.

On the wind. A vessel is said to be sailing on the wind or by the wind when the wind comes from a direction forward of the beam.

Points of failure. Areas that are going to be the cause of a catastrophic failure if allowed to progress to their terminal end.

Polyester. Plastic fibers that is used to make ropes.
    Double braid. Rope that is made out of a core and a cover. Dependent on the core for strength.
    Three strand. Rope that is made by twisting three strands of fiber to make a rope. Resulting rope has a high degree of elasticity and stretch, making it ideal for applications where a little give is welcome.

Preventer. Running rigging used to pull forward on the boom to prevent an accidental jibe.

Pythagorean Theorem. $a^2+b^2=c^2$. The relationship between the base and height of a right triangle and the hypotenuse.

Reach. To sail broadside to the wind.

Reefing. The act of making sails smaller to reduce their power, usually performed as winds build in strength.
    Reefing points. Specific areas of the sail where hardware is connected to facilitate the act of making the sail area smaller.
    Reefing lines. Running rigging used to manipulate the sail and set it in a reefed position.

Rig inspection. The act of actively looking over and assessing the condition of the rigging on a sailboat. This is a critical task that identifies problem areas before catastrophic failure occurs.

GLOSSARY

Rig measurement. IPEJ, where I is the height of the headsail, P is the length of the mainsail luff, E is the length of the boom, and J is the length of the deck between the mast and the headstay.

Rigging. Collective components used to support the sails in the breeze to drive a sailboat forward through the water.

Righting moment. The tendency of a sailboat to return to an upright position when heeled over.

RM30. Righting Moment at 30°. Represented by the amount of force needed to heel the boat over 30° and this value is used to calculate all the rigging components.

Roach. The convex curve to the leech of a sail enabled by having battens in the sail.

Running backstay. Stays that are set and doused to resist the forward pull of the staysail on the inner forestay. They interfere with the boom, so they need to be managed to allow space for the boom to swing through.

Running rigging. Ropes that are used in raising, lowering, and trimming sails.

Safety factor. Value that exceeds the minimum required strength for a component.

Sail. A curved surface used to transfer wind speed into the driving force for a sailboat.

Sail balance. Trimming the sails to bring the CE over the CLR, resulting in a balanced helm.

Schooner. A fore-and-aft rigged vessel with the larger or mainmast aft.

Sheet. The rope used to trim a sail.

Sheet block. Block used to guide the sheet from the clew to the crew to trim the sail.

Sheeting angle. Angle between the tack and the sheet block relative to the midline of the boat.

Shrouds. The parts of the standing rigging that support the mast laterally, extending from the masthead or tangs to the chain plates of the vessel.
   Cap Shroud. Shroud that runs from the chainplate to the top of the mast.
   Intermediate Shroud. Shroud that runs from the chainplate to the base of the second, and successive, spreaders.

GLOSSARY

Lower Shroud. Shroud that runs from the chainplate to the base of the first spreader.

Shroud frapping knot. Knot invented by Dr. Harry H. Benavent to secure the lashings used to tension synthetic standing rigging, thus allowing the stay to be tightened and maintained tight while the shroud frapping knot is tied, holding the tension in the stay via the friction between the frapping turns of the knot and the sides of the lashing line.

Sloop. A similar rig to the cutter except that the mast is farther forward.

Snubber. High-stretch rope used to add elasticity to the ground tackle and anchor rode. Also reduces stress on the anchor raising equipment by transferring the load to a different location. Reduces shock loads, which help the anchor remain firmly seated in the seabed.

Spinnaker. Large triangular sail used for running and broad reaching. Sail is set flying and is attached to the boat at three points.

Splice. Alternative to knotting where the fibers of the rope are woven into another rope to connect the two, usually resulting in a strong connection without any loss of strength of the rope.

Spreader. Device used to hold a shroud away from a spar to improve the angle at which it connects. This increases the effectiveness of the rigging and allows for taller rigs.

Stability. The lack of unexpected movement from a sailboat.

Stainless steel. A steel alloy that has less carbon and more chromium allowing it to form a protective oxide layer that delays corrosion. Susceptible to crevice corrosion. Common alloys used on sailboats are 18-8 for fasteners, as well as 304, 316, and 316L for components.

Standing rigging. Shrouds and stays, usually wire, used to support the masts and spars.

Stays. The parts of the standing rigging that support the masts in a fore-and-aft direction.

Staysail. Sail usually set on the inner forestay, but technically can be set on any other stay that is not the headstay.

Storm sail. Smaller specialty sails made out of heavyweight canvas for the express purpose of being used in severe weather.

Storm sheet. Running rigging used to trim the storm sails; significantly stronger than the working sheets.

Tack. The lower forward corner of a sail.

Tacking. The operation of sailing a vessel against the wind on a zigzag course.

Toggle. Small component that is used at the end of a stay to add an axis of rotation, turning a connection into a universal joint, reducing the amount of stress and strain on the terminators of the rigging.

Topping lift. Piece of running rigging that is used to prevent the boom or other spar from falling when the sail is not raised.

Traveler. Running rigging used to trim the mainsail and adjust the position of the mainsheet block.

Triatic stay. Standing rigging that runs from masthead to masthead on a multimast vessel.

Trysail. Storm sail used in lieu of the mainsail that does not involve the boom.

Turnbuckle. Mechanical device used to adjust the tension in standing rigging.

Vang. Running rigging used to prevent the boom from rising when the mainsheet is eased and no longer has reduced efficacy when on a broad reach or run.

Weather helm. Sailboat's tendency to turn upwind if the rudder is maintained straight.

Weight aloft. As weight is raised higher above the center of gravity of the boat, the torquing force that it exerts is increased. Ballast is needed to counteract this weight to keep the boat upright. The weight of the rig as well as all the associated components that are attached to the rig contribute to this weight. This weight has a direct and negative effect on the performance characteristics of the boat.

Wire rope. Rope made out of wire, typically 7×7 wire is used as this has good strength as well as good flexibility.

GLOSSARY

Wood. Natural material used to manufacture various components on a boat. Spars are ideally made out of Sitka spruce but other clear grain lightweight and high-strength woods will also serve the purpose.

Working sails. Usual set of sails for a sailboat, not the specialty sails such as light air or storm sails.

Index

age of discovery, 23
aluminum mast, 37
anchoring
   anchor rode, 231
   snubber, 232, 272–273

barque, barc, or bark, 24
barquentine, 25
bend in mast, 48
brig, 24
brigantine, 24
bronze, 99, 100, 106

canting mast, 53
carbon fiber mast, 39
Center of Effort (CE), 75, 82, 85
Center of Lateral Resistance (CLR), 75, 82
chafe, 177–178
chainplate, 97, 106–111
   bolt, 111
   bronze, 106
   carbon fiber, 110
   stainless steel, 107–108
   titanium, 109
cleat size, 230
clevis pin, 106–110, 285–286
clipper, 25
coefficient of thermal expansion, 37, 38, 39, 138
composite chainplate, 104, 110
compression fittings, 134

continuous rig, 9
corrosion, 165–168
crab claw sail, 21
cracks, 171–174
crevice corrosion, 102, 169–171
Cunningham, 17, 56, 222, 268
cutter, 25, 26, 83, 84, 117, 217, 218, 251, 255, 292

deck step, 40, 46, 94, 139, 216, 242
diagonal stay, 9
diamond stay, 9
dock line, 227–230, 271
double reef line, 19, 67
downhaul, 16, 72, 226, 259–260
draft, 14, 16, 56
drifter, 90, 266
Dyneema, 135–137, 145–161, 177–183, 243–249
   eye splice dimensions, 244
   splicing, 145–152

fisherman sail, 89
folding rule method, 287–288
foot pound, 31
fractional rig, 93
fully rigged ship, 25
furling line, 70, 227, 269–271

galvanized wire, 114, 115
genoa, 82
golden age of sailing, 23

INDEX

gollywobbler, 89
guys, 17

halyard, 14, 56, 252–256, 277–281
    replacement, 277–281
haul, 15, 62, 63
historical sail variations, 21–24
hull speed, 130

inversion, 141, 216, 284, 294, 295

jackline, 18, 232, 233
jib, 80, 81,

keel step, 40, 139, 212, 215, 242
ketch, 28

lazy jack, 18
lifeline, 18
lift, 15, 59, 256
light air sails, 91
linked rig, 9
luff tension, 14

mainsheet, 258–262
mast, 31, 50
    aluminum, 37
    bend in mast, 48
    carbon fiber, 39
    wooden, 33
mast connection, 238–240
mast step, 40

outhaul, 15, 58, 225–226

pound feet, 31
preventer, 273

reef line, 19, 63, 69, 224–225, 257–259
    double reef line, 19, 67
    material, 224
    points, 19, 6364

setup, 69
single reef line, 19, 64
remaining life in components, 195–196
rigging, 1
    load, 128
    inspection, 185–189
    inspection checklist, 190–193
    size, 133, 285–289
    tension, 287–289, 139–142
    tune, 283–295
RM30, 121–126
rod rigging, 116
rotating mast, 53
running rigging, 13–19, 275–277, 251–274
    halyards, *see halyard above*
    reef line, *see reef line above*
    replacement, 196, 199–204
    sheets, 17, 60, 221, 226–227, 260–268
    sizing, 275–277

sail
    fisherman, 89–90
    genoa, 82–83
    gollywobbler, 89–90
    jib, 80–81
    mainsail, 56, 59–60
    mizzen, 15, 24, 29, 305
    staysail, 83–87
    storm, 226
    trim, 14, 16, 56, 60, 77
    working, 91
schooner, 26, 27
sheet, 17, 60, 62, 221, 226–227, 260–268
shroud (cap and lower), 8
single reef line, 19, 64
sloop, 25, 26
slutter, 27
snubber, 232, 272
solent, 27
spinnaker, 87, 266
spreaders, 43, 49, 139–142
stainless steel, 100, 101, 107, 108

INDEX

standing rigging
   backstay, 7
   backstay tension, 94–95
   cap shroud, 8
   headstay, 7
   lower shroud, 8
stay (backstay and headstay), 7
staysail, 83, 84, 266–267
stop knot, 274–275
storm sails, 91
studding sail, 90
swage, 133–134
synthetic standing rigging, 116–120, 139–142, 145–161, 211–219, 243–249

tack horn, 68
titanium, 103, 109
topping lift, 256
traveler, 18, 223, 269
trysail, 77, 78, 267–268

vang, 17, 268

wire strength, 285–286
wooden mast, 33
working sails, 91

yankee, 83
yawl, 28

www.ingramcontent.com/pod-product-compliance
Lightning Source LLC
LaVergne TN
LVHW081529060526
838200LV00045B/2046